Low Power Design with High-Level Power Estimation and Power-Aware Synthesis

T0189520

Sumit Ahuja • Avinash Lakshminarayana
Sandeep Kumar Shukla

Low Power Design with High-Level Power Estimation and Power-Aware Synthesis

 Springer

Sumit Ahuja
ECE Department
Virginia Polytechnic Institute and State
302 Whittmore Hall
Blacksburg, VA 24061
USA
sahuja@vt.edu

Avinash Lakshminarayana
ECE Department
Virginia Tech
302 Whittmore Hall
Blacksburg, VA 24061
USA
avinashl@vt.edu

Sandeep Kumar Shukla
Bradley Department of Electrical and
Computer Engineering
Virginia Tech
302 Whittemore Hall
Blacksburg, VA 24061
USA
shukla@vt.edu

ISBN 978-1-4899-8780-8   ISBN 978-1-4614-0872-7 (eBook)
DOI 10.1007/978-1-4614-0872-7
Springer New York Dordrecht Heidelberg London

Printed on acid-free paper

Springer is part of Springer Science+Business Media (www.springer.com)

To the friends and families,

and all our fellow Hokies.

– Sumit Ahuja
and
Avinash Lakshminarayana
and
Sandeep K. Shukla

Preface

Designing low-power computing hardware has always been a priority since the early 1990s. The famous graph drawn by Intel's Shekhar Borkar is now imprinted in the minds of designers. It was realized that as the clock speed is scaled from mega hertz to giga hertz, the heat density on the surface of silicon chip would compare with that of rocket nozzles to the surface of the sun. Computing in today's era has become pervasive in the form of handheld devices, smart phones, tablet computers, and most importantly bio implantable devices. Borker's graph aptiy captures the need for reducing heat dissipation subsequently battery life conservation. The battery life of devices that are implanted inside one's body must be sufficient to not require surgical substitution every few years. The wireless sensor network technology deployed for reconnaissance purposes by the military or for disaster management scenarios also brought in the requirement of long battery life despite the energy expensive communication functionalities.

In the past few years, the mode of computation has changed so rapidly that ubiquitous and pervasive computing is a reality. However, this wide spread use of computing and communication in our everyday lives, and for critical applications, the issues of privacy and security surfaced significantly. This meant the ability to encrypt, and decrypt large amounts of information fast and with as little power expense as possible. The usage of wireless communication of multi-media data over the next generation of wireless technologies demanded energy-efficient and fast compression/decompression and coding/decoding hardware.

Finally, to accelerate the graphics performance, and cryptographic or data compression, the wide spread use of accelerator hardware in the form of co-processors also demanded low-power implementation of such co-processors. The software implementation of these functionalities just did not make the required performance, and even in overall energy usage, the software implementations started getting worse.

The confluence of all these technological trends, and a few others, the low-power hardware design, system level power management, and power minimizing optimization of applications through compiler techniques, and even programmatic techniques through power aware system APIs took off as research areas in the middle of the 1990s. The quest is still on and with a vengeance because the battle for low-power design is still very much on.

In the mean time, the power efficiency crisis has led to other developments such as multi-core system-on-chips which operate at lower clock frequency but make up for performance by virtue or the extra opportunities for parallel execution. Also, the introduction of graphics processor based accelerated computing, heterogeneity in the form of co-processor based acceleration, renewed vigorous interest in parallel programming models, parallelizing compilation, etc., are all the side benefits of the power envelope that posed as a threat to Moore's law.

However, even with the parallelism based performance enhancement the goal of reducing power consumption remains – the cores or the graphics processors or the co-processors should use as little power as possible. However, keeping the performance within the power and temperature envelope so as to avoid using cooling mechanisms would make the system bulky and expensive.

Our foray into low power design came about 10 years ago. One of the authors while working at Intel in 1999, had the good fortune of modeling hardware blocks with the first versions of SystemC (-0.9), and Cynlib, and had the first introduction to using C/C++ to hardware design. While building the simulation model, an interesting observation to note was the same techniques one employs to reduce the simulation time in cycle based simulation can also be used as a power saving feature in the actual hardware. For example, clock gating is known to reduce power consumption, and in cycle based software simulation of the hardware, identifying the conditions when some variables do not need to be updated in each cycle would reduce the computation in each cycle's simulation. Similarly, using bit vectors for computing instead of computing each individual bits during simulation would reduce simulation time of a cycle, and also with suitable instructions for changing multiple bits at the same time using bit vectors would reduce power by abstracting the logic of each individual bit's computation into a single computation. These techniques are obviously not always power saving (clock gating may incur power consumption if the opportunity does not arise during the computation, and bit vector operation may not reduce power because each bit's computation may be drastically different from other bits and hence lack of abstraction of the overall logic for the entire vector might incur more power consumption). However, heuristically these work well for specific instances, and identifying those instances itself is an interesting research problem – potent with possible usage of formal methods in power reduction.

In 2001, again the work in dynamic power management for system level power reduction interested one of the authors, and led to statistical techniques for predicting the inter-arrival times of inputs by fitting probability distributions from past data. Again formal methods based in probabilistic model checking turned out to be helpful in this regard, and subsumed many past work based on Markov chain and Markov decision processes within the one generic methodology. This led us to believe more in formal methods based approaches in abstracting many apparently disparate approaches into a single methodology. The other interesting aspect of these was reuse of already existing formal methods and tools that when properly used actually simulated other methods which were algorithmic in nature, but did not yield themselves to easy analysis.

In 2004, we had a very interesting problem to look at. Research in high level synthesis to generate register transfer level (RTL) hardware from behavioral descriptions suddenly gained popularity again after a few years of hiatus when Synopsys had abandoned its behavioral compiler. This was necessitated by the surge of hope in the community to use system level design languages to describe entire hardware/software systems. These languages help in simulating these models faster to debug the correctness issues, and then the same model can be used as a reference model from which hardware software partitioning and actual implementation are done. The term ESL (Electronic System Level) became popular, and the design automation conferences such as DAC, DATE, ASP-DAC – all started focussing on ESL based design methodologies. It was clear to everyone that maintaining an ESL model of the system, and manually implementing RTL may not work very well. The reason is because at RTL designers diverge from the features modeled in the ESL model, reapplying those changes to the ESL model which was used for initial debugging of the functionalities is cumbersome, and often not done by engineers. That implied that unless synthesis of hardware is done from such high level model, and any changes to be applied are first applied to the high level models and the lower level models are re-synthesized, the two models will quickly diverge, rendering the advantages of high level modeling somewhat moot. Thus, a surge in high level synthesis research and also industrial tools started appearing. Forte Design System's Cynthesizer came about as a tool for synthesizing RTL implementations from SystemC models. However, SystemC being implemented as a class library in C++, and C++ being not strictly formally defined left room for ambiguity. And for the same reason SystemC could not be used to generate equivalent hardware. The other problem that still challenges synthesis solutions for SystemC to RTL path is that the concurrency model of SystemC is an afterthought rather than integral part of the language. Use of cooperative threading model in SystemC is good enough for simulation, but it does not express the parallel activities that happens in a real digital hardware adequately. Also, the designer has to schedule the synchronization between threads at communication points, and cycle boundaries which is always error prone – as evidenced by the plethora of bugs that are seen in multi-threaded software programming.

As a result, alternative models of computation to express the designs in SystemC such that the model of computation captures exactly how the intended hardware would behave, interested us immensely. However, in order for these models of computations (MoC) to reflect the behavior of the real hardware, one could take two different approaches. In one approach, create class libraries that encapsulate the model of computation, and then instantiate the hardware as an object of that specific class. However, while for synthesis purposes this may as well work, the simulation would be extremely sluggish because one would have to simulate using the discrete event simulation kernel that came with SystemC reference implementations. While some models of computations are quite compatible with such simulation, most are not – for example, dataflow computation can be simulated much faster if discrete event simulation is abandoned in favor of its own native simulation kernel. Thus, we implemented a multi-MoC simulation kernel for SystemC, and tried to convince the community that it is the right path to take. While academically this was great success measured by the number of publications, but in reality, the SystemC consortium or Accelera committee for SystemC standardization did not pay much attention to this line of thought. These committees kept the discrete event kernel alive even today while introducing abstractions after abstractions which include transaction level modeling. However, we have told that story in two books, and in many conference papers, journal articles, and book chapters.

Nevertheless, one company that has been partially successful in changing the model of computation that is more in line with how the hardware behaves rather than how it simulates is Bluespec Inc. When we say partially successful, we mean that despite its suitability in modeling hardware, and a very effective synthesis tool, not a large fraction of the system houses actually use this tool yet. Again, this is possibly because of unfamiliarity of designers with a distinct model of computation, even though it would reduce their concurrency bugs dramatically. We conjecture that this is due to the fact that most hardware designers got trained with languages like Verilog and VHDL which were invented as simulation languages with discrete event simulation at their core. In any case, the point is that Bluespec introduced a way to model concurrent actions in an action oriented or rule oriented fashion, such that all the actions are atomic, and among such atomic actions, the ones that do not conflict with each other can be scheduled in a clock cycle. This scheduling is done by the Bluespec compiler. However, the original Bluespec compiler scheduled all possible non-conflicting actions in a cycle to maximize performance – but it could increase the peak power – challenging the temperature envelope.

So we ceased the opportunity to optimize the scheduling technique using bi-criteria optimization algorithms to make the compiler schedule within the peak power bound. This led to certain complications in terms of correctness proofs. The Bluespec model being a collection of concurrent rules, its semantics is nondeterministic. Any possible sequential scheduling of these rules constitutes a valid behavior, and hence any implementation whose behavior is included in these valid behaviors is a valid implementation. Thus, the original scheduler that implements one of the behaviors would show a different behavior than our peak-power optimizing scheduler induced behavior. This means that we must formally prove that our optimizing scheduler indeed induces an implementation of the original specification even though it does not exactly match cycle by cycle with the one produced by the original compiler. This problem was solved by using formal verification method with automata based behavior inclusion proof techniques. However, this point illustrated at the same time the strength and weakness of Bluespec. The strength of Bluespec being that it allows multiple possible implementation of the same specification – which means it does not overly constrain the specification which necessarily happens when one designs directly in Verilog and VHDL. So if the designer makes poor implementation choices, those choices become part of the design. Bluespec allows the compiler to make the decisions regarding implementation. The weakness is not that of Bluespec per se, but that of designers. Standard hardware designers – trained to design deterministic hardware in Verilog/VHDL by imposing their design choices which are not necessarily the only possible choices to implement the required functionality – find it hard to accept that their specification leaves open a plethora of possibilities. This makes them unsure of what their specification accomplishes – hence, they over constrain their Bluespec specification so that eventually the strength described above vanishes, and the specification becomes the implementation. This means that the choices one could make via a power optimizing compiler are no longer existent. This point was belabored by us initially, and but eventually we conceded that it is not the problem of the language or the tools – but that of designers who are exposed to over constrained RTL design.

At the same time, we also worked on introducing operand isolation, and clock gating features during synthesis process from Bluespec specification to implementation – these have been duly described in another book. The reason we bring out the entire history of our foray into low power hardware design is because our own history parallels also the history of the progress in the field to some extent. So we think it is an interesting read for the readers of this book to put the content of this book into a broader chronological perspective. Now we come to the work reported in this book. This work was mostly carried out by two of the authors as part of their Doctoral dissertation and Masters thesis work. All the work reported here some how pertains to power reduction techniques applied to as high level as possible, be it for co-processors, FPGA implementations of functions, or ASICS, and our slant has always been to reuse techniques from formal verification and possibly use some of the tools that already exist in that domain.

There are two problems that are necessary to be solved for any methodology that aims at reducing power consumption, especially if the methodology is automated. One has to be able to estimate the power consumption as accurately as possible from the high level models themselves without having to synthesize all the way to hardware and then use well established estimation tools at that level. This question has been plaguing the practice of system level modeling from the very inception. In the very beginning – back in 1999 – a lot of industrial engineers dismissed the use of high level modeling showing this exact problem as the cause. Indeed, if your model does not have enough implementation specific details, how could you estimate timing and power and other parameters that decide the quality of the choices made in the design. If one could estimate these numbers from system level models themselves, then the architectural exploration would be always done at the system level. The skeptics of course had a point – how would anybody know the physical aspects of a design when it was just functionally specified (or implementation independent) at such a high-level?

To answer this questions one could make two points. First, if we can show that the relative power consumption between two alternative design specification (for example, in one you would use memory and in another a suitable number of registers; or in one case you would like to use a co-processor and in another you would implement a function in software) can be shown to reflect the same degree of relative standing in the implementation irrespective of the implementation details and technology – we can invent ways to profile the high level model in some way, and generate the relative numbers. Second, we do not answer the question just for the high level model it self, but the answer is always given for a pair – the model and a implementation choice set. In this case, your numbers are not specific to the high level model, but specific to the implementation choices paired with the model.

The first possibility is easily proven to be impossible. For example, if you have two models for the same functionality, but with distinct architectural choices, D1 and D2. The first case demands that if p() is an estimation function at the high level independent of any implementation knowledge, and if the implementations of these two design choices are I1 and I2 with the same technology and other lower level configuration choices, and if e() be the estimation function at that level, then p(D1) < p(D2) must imply e(D1) < e(D2) for all design choice pairs and corresponding implementation choices. An extremely tall order, and one can easily come up with examples where this does not hold no matter how clever the estimation function p() is conjured up.

Thus, we did not even attempt to come up with estimation functions that are totally independent of low level choices that would be made. However, if we can export some of the information from the low level choices from suitably controlled experiments, and use that information within the estimation process at the high level, we can obtain quite accurate estimates or at least relative estimates that are useful in making architectural exploration. Thus, if C_i(for $i = 1,\ldots,n$) are the choices of low level implementation decisions (such as whether asic, or fpga implementation will be done, what technology node will be used, what technology libraries will be used, what kind of memory technology will be used, etc.), and for the same functionality we have $D\_j(j = 1,\ldots,m)$ architectural choices at the high level, then our estimation function e() will now take two arguments – $p(D\_j,C\_i)$ will provide us with an estimate of average power consumption that would result from choosing the high level architectural features f of D_j and the implementation choice C_i. Note that real power consumption when this functionality is actually implemented with these choices will be $p * (D\_j,C_i)$, and we want to make sure that $|p(D\_j,C\_i) - p * (D\_j,C\_i)|$ to be bounded or the percentage of estimation error $100 * |p(D\_j,C\_i) - p * (D\_j,C\_i)|/p * (D\_j,C\_i)$ is small enough to be tolerable.

A suitable estimation methodology leads to a good architectural exploration technique – especially when comparing multiple different architectural variations at the system level, and multiple target technologies such as ASIC, FPGA or software.

In the power estimation field, especially at the register transfer level, macro-modeling is a common technique that has been used for creating simple models that capture the power consumption characteristics by parts of a design, and during simulation time, those macros can be used to substitute the real design – leading to faster simulation, but a faster estimation also for full chip power consumption estimation.

However, macro-modeling is quite bit of an art than engineering – as one has to capture the various elements in the design that are typical sources of power consumption for that design. For example, one could consider the toggle count of signals, and registers, because the toggles indicate capacitive charging and discharging and that is a source of power consumption. Also, the frequency of the clock that drives the hardware is a prime indicator. Along with that, the area of the component plays a significant role in power consumption. In fact, a crude power model is that the power consumption is proportional to the area, the clock frequency, and the square of the voltage level at which the hardware operates. Thus, reducing voltage by voltage scaling technique is a well known technique for power reduction provided the lower voltage does not affect the correctness of the computation which is a legitimate concern because lower voltage reduces the noise margin. Similarly, frequency scaling is another way to reduce power, but one has to make sure that the scaling of the clock speed does not unbearably affect the performance of the design. Reducing the area is technology dependent, but there is also the toggle or activity factor that could be reduced by techniques such as operand isolation, clock gating, etc.

A crude macro-model which is only dependent on area, frequency, toggle count, and voltage level does not always provide the most accurate power consumption estimate. This is because the toggle count is often inaccurate because it is usually estimated based on input and output toggle as the internal signals may not be easy to observe, or instrumentation of the model to get accurate toggle count may prove to be expensive. So the best way to go about that is to find other ways to estimate the toggle count and also to capture the effect of the toggle count . For example, depending on which kind of gate is toggling and other factors, the power consumption may vary. Also, viewing a digital design as a state machine, and estimating the power consumption characteristics at each of the states, and finding out what percentage of the time, the design stays at which state could provide a basis for a macromodel.

Also, recall that there are different ways power is consumed by hardware. The dynamic power vs. static power is one way to categorize them. The dynamic power is what we have discussed earlier – the power consumed during operation of the hardware is considered. The static power is consumed due to transistor leakage which is getting considerably high as we move from 45 nm to 22 nm and beyond. In this book, we did not consider static power consumption issues directly. However, some of the low level tools we used to create our database which was then analyzed to create our power models did estimate leakage power or static power.

We also consider the notion of peak power which is the maximum instantaneous power consumption – this may not directly affect the overall energy usage because one could have low peak but run the clock slower, and hence take more time – and more energy even with a lower peak power. However, peak power does indicate thermal characteristics of the hardware, and hence often needs to be reduced. In this book, peak power is also not given as much consideration as average dynamic power consumption.

Coming back to macro-modeling – one approach is statistical. In this approach one establishes a correlation between multiple observed measures and the overall power consumption. This requires two phases: in the first phase we decide on the variables that we want to measure – as we establish good analytical reason to believe that they affect the power consumption. We then consider a number of high level designs (could be algorithmic C, SystemC transaction level model, etc.), and the corresponding hardware implementation at the RTL level. We execute the same test cases on both the high level model and the RTL level implementation for each design, and measure the variables while doing so at the high level. We then use an RTL level power estimation tool such as PowerTheater to estimate power consumption for those test cases. After sufficiently large number of such experiments we have enough sample from which parameter estimation for a regression model can be done. In the second phase, we use this data and least square estimation technique to establish a linear relation between the measured variables and the estimated power consumption at the RTL level. Under certain statistical tests, if the estimation is proven to be sufficiently accurate, we then use this linear model as the macro-model for power consumption on any similar hardware blocks. The accuracy of estimating power consumption of a design which is available only at the high level turned out to be quite good by using this method.

The question then is what kind of variables of a design do we need to measure for the above technique to be successful. One could just consider area of the design as a variable. One could also use the different types of elements in the design (combinational and sequential elements). One also could consider input signal toggles or output signal toggles. One technique that proved quite useful is when the high level design can be looked at as a finite state machine with data path (FSMD). One could consider the toggles of signals at the different states as distinct variables, and the use those variables in the regression. This gives a finer grain nuance to the correlation because even if two states have the same number of toggles they affect power consumption differently. For example, if the design has a low power state and a high power state, chances are the toggles at the low power state will consume less power than the toggles in a high power state. Thus, distinguishing the toggles at different states and distinguishing their effects on the overall power consumption by weighing them by the fraction of the time the design stays in the corresponding states, and estimating the coefficients that indicate the influence of the states on the overall power consumption, we obtained linear power models with much higher accuracy than done otherwise.

The regression based technique also turned out to be useful when considering FPGAs as target platforms. However, in this case, it turned out that one could classify various IPs into a number of classes, and parameterize the linear regression function so that for different classes of IPs the regression functions vary. As explained earlier, the estimation is only one piece of the puzzle. The other part is power optimizing refinement steps from high level models to low level implementation. Traditionally, savvy engineers do this by virtue of experience and intuition. They change the architecture, algorithms, and other properties of the design to lower power consumption. Sometimes intuitions may work, and estimation followed by the proper modifications of the architecture or algorithm could give the designer some idea about if the tricks used in the design to lower power indeed works. However, in the absence of such engineering experience, one could use a tool that takes a higher level description, and synthesizes RTL that implements the design.

However, a power optimizing synthesis or compilation will also insert logical precepts that would reduce power consumption – such as clock gating, operand isolation, etc.

We got the opportunity to do that in our work in collaboration with Cebatech Inc, which is a IP company but also has an internally developed tool that synthesizes competitive quality RTL from algorithmic C models. Finding clock gating opportunities from the high level model by analyzing traces of simulation or by analyzing the code and then automatically inserting clock gating logic in the appropriate places is challenging – especially because we only want to insert clock gating if it indeed decreases power consumption. This means that registers which do not change its value often but are on active paths more often should be identified. There could be two types of opportunities to do clock gating: one we call combinational clock gating and another we call sequential clock gating. Combinational clock gating is based on a logic that computes an enable signal based on the signal values in the current cycle. Sequential clock gating allows one to compute an enable signal based on signal values in past cycles, and depending on how much buffering of past values you can afford within your area and power budget, you could use values from deep past, if there is enough evidence that the deep past values can determine whether a register needs to be clock gated in the current cycle.

In a related work, we have shown that this computation of the logic based on past values of the signal is actually a fix point computation, and the requirements can be captured with modal mu-calculus, and fixed point computation may give you all possible past influences on a register in the current cycle. However, this fixed point computation may turn out to be quite expensive, and hence one could approximate the fixed point, and that captures you at least some influences until some past cycle. In this book, however, we have not discussed this particular theoretical result, but it may be found in [36] paper. We mention this paper in particular because in the recent times multiple algorithms have been proposed for sequential clock gating which basically does influence analysis through time unrolling of the design, and using graph algorithms to identify influences. One could also use backward constraint propagation technique to identify finite depth of past influence. However, the fixed point based formulation of the problem basically generalizes the entire idea of computing sequential clock gating logic, and explain in clear terms what each of those papers actually are heuristically attempting to do.

Another study related to the clock gating we did is the granularity of clock gating. A bit by bit clock gating vs. word level or byte level, etc., often allows us to reduce the gating logic but it depends on the specific logic on which this is being done. Thorough comparisons of these are essential to understand the impacts of clock gating in power reduction.

Using our methodology to enable clock gating logic from the high level model itself and using that information during RTL synthesis to obtain already clock gated RTL, we have been able to carry out multiple architecture exploration projects for power reduction in co-processors. Various non-trivial IPs at Cebatech including compression and decompression co-processors were designed for low power and high performance through these exploration techniques at a high level.

During the initial phases of this work, our goal was to minimize the designers efforts for adding power consumption lowering features into the design. We therefore, looked into reusing some of the analysis of a design that was done during formal verification phases of the design cycle. This approach was illustrated with a design environment from Esterel technologies which was called Esterel Studio. Esterel Studio allows one to design hardware and software system in a formal language of synchronous programming paradigm – in particular, Esterel language. One could also write formal assertions in Esterel. The tool had the abilities to synthesize RTL hardware from such specification, and also use the assertions in formal verification, and assertion checking during simulation. The assertions can be written to show reachability of certain specific states. In this work, simulation trace was used to construct modes in the design (Modes are macro-level states which the execution goes through). Modes could be so constructed so that each mode has a different power signature based on the combinational activities in each mode. Reachability in each mode is tested through assertion checking. Also, to collect activity information in each node, assertions were used to construct test cases that would take the design into the required mode. The simulation dump for these tests can then be analyzed to construct the activity factors, and thus estimation of power consumption. This is an interesting approach; however, Esterel technologies did not make a lot of traction in the hardware design arena most possibly due to the designers' unwillingness to accept a completely new paradigm of design entry language.

This preface does not cover the content of all the chapters of this book. However, it outlines the context and mindset with which this research was carried out, and how the specific historical contexts shaped the approaches we took in solving the problem of power estimation at the system level, and inserting power reducing constructs during high level synthesis.

Obviously, these are not the only possible methods, nor does this book claim to be exhaustive in referencing related work. However, we believe that the content of this book could allow a reader to familiarize him/herself to recognize the problems; some possible solution approaches that worked at least as far as simulation based studies are concerned. As a result, while not a definitive handbook on power estimation at high level or power reduction during synthesis, it certainly will serve as a book that could help those uninitiated to the field of low-power design at the high-level or system level.

Acknowledgments

We acknowledge the support received from NSF PECASE, NSF-CRCD, and Cebatech Inc. grants, which provided funding for the work reported in this book. We would like to acknowledge various people with whom we were directly/indirectly associated with during the duration this work is performed.

Sumit would like to thank his parents Ram Swaroop Ahuja, Uma Ahuja, his family members Vishal, Amita and Kashvi for their continued love, support, and encouragement in his endeavors, and for always being there for him. Special thanks to his wife Surabhi for motivating him to complete a lot of work for the book after office hours. Sumit would like to thank his friends Luv Kothari and Deepak Mathaikutty for motivating and guiding him during the course of PhD. To Wei Zhang for numerous hours of technical discussions and guidance during my PhD, and for being patient and supportive through those discussions. To Gaurav Singh for helping and discussing many research topics related to this thesis. To all his roommate, friends, colleagues from Virginia Tech and friends in bay area: Mainak Banga, Hiren Patel, Syed Suhaib, Nitin Shukla, Debayan Bhaduri, Bijoy Jose, Harini Jagadeesan, Mahesh Nanjundappa, Bin Xue, Anupam Srivastava and Ananya Ojha. To all the other friends that have come and gone – thanks to all.

Contents

Chapter 1
Introduction

Managing power consumption in complex System-on-Chips (SoCs) and custom processors is emerging as the biggest concern for designers. These SoCs and processors are vital parts of embedded systems used in electronic equipments ranging from laptop computers, mobile phones, to personal digital assistants, etc. Quality/Performance of such devices is not only measured by what kind of functionalities these devices are capable of performing, but also how long these gadgets survive without plugging into power outlets for recharging. There is an increased demand for reduced form factor or lower size of such devices, which is generally determined by the size of the battery used to run these devices. Another need for controlling power consumption of the design is to decrease power density of devices on a single die. A survey report [1] shows that power density of some designs can reach the power density of a rocket nozzle in future, if power aware design approaches are not used. The reasons cited above have motivated engineers to look into power aware design methodologies.

Traditionally, the hardware design flow starts from the register transfer level (RTL) whereupon the design is synthesized to the gate-level, and after further processing at the lower abstraction levels a chip is manufactured. Methodologies for power estimation and optimization at the RTL or lower-level are well studied. In the recent past, modeling at higher level of abstraction was advocated for quick turn around. Increased demand and reduced time-to-market requirements for electronic equipments have forced designers to adopt high-level modeling in their design-flow. Design methodologies, where reduction of power consumption of such systems can be done at the high-level, have become the need of the hour. We know that any optimization requires good analysis of the design, and so is the case with managing the power consumption of the design. To reduce the power consumption in the final product, it is necessary to have an estimate of the power consumption early on. It is becoming essential to utilize design methodologies to reduce and estimate power consumption at the high-level.

S. Ahuja et al., *Low Power Design with High-Level Power Estimation and Power-Aware Synthesis*, DOI 10.1007/978-1-4614-0872-7_1,
© Springer Science+Business Media, LLC 2012

1.1   Trends in System Design with System Level Modeling and High Level Synthesis

Advances in semiconductor technology can be well represented by *Moore's law* [109], which states that every 18 months or so, almost double the number of transistors can be accommodated on the same die size.[1] Because of such advances, every new generation of microelectronic systems can accommodate more computing elements than before. With the increased availability of computing elements, more and more applications are mapped onto the hardware. Increased demand for hand held electronic equipments capable of utilizing more applications require increased focus on system-on-chip (SoC) design approach. In an SoC design, a lot of reusable hardware blocks or intellectual properties (IPs) are integrated together. *Electronic System Level (ESL)* based design approaches are becoming necessary to design complex SoCs.

ESL methodologies are mainly targeted to improve the productivity of the designer. These methodologies help in reducing the design cycle time of the hardware design by adopting approaches that can be applied at the higher-levels. Some of the reasons why ESL is getting traction are the need for fast prototype development, quick functional verification, and easy architectural exploration. Most of these system models at such high-levels are very abstract and many details of the real hardware are not available. This level often lacks crucial information such as pipelining required to achieve appropriate throughput, selection of appropriate hardware blocks to perform the desired computation such as choice of various multipliers, finite impulse response (FIR) filter, etc. efficiently come up with solutions which are comparable to manual designs.

High-level models are generally written by algorithm/system developers in C/C++. Traditionally, once these algorithms are fixed, hardware design starts from scratch using Verilog/VHDL. Hardware generation from the algorithmic description of the design is still more or less an unachieved goal. An algorithm developer generally thinks about the execution of a model using C/C++, which is sequential by nature; while hardware is inherently parallel. The gap between software implementation stage (algorithmic description) and synthesizable hardware implementation (RTL description) is often very large. In traditional design methodologies, hardware modeling at the RTL requires effort ranging from several months to a few years, depending upon the size and complexity of the hardware designs.

---

[1]More recently, this period has roughly become 2 years.

1.1.1   The Need for High Level Synthesis

High-level specifications allow the designer to ignore various RTL coding issues (such as synchronization and scheduling of operations) while capturing the functionalities of the target design. As stated above, High-level to RTL conversion is very tricky because each algorithmic functionality can have multiple implementations. Modeling of hardware designs at high-level has gained wide acceptance in functional verification, generation of quick simulation models and virtual prototyping. Various efforts to shorten the time consuming transformation of algorithmic description to RTL are addressed by the High-Level Synthesis (HLS) techniques [28,42,101]. Most of the high-level synthesis solutions convert the high-level specification model to the implementation model (such as RTL model). There are various levels of abstraction on which high-level specifications can be captured starting from the transaction level modeling (TLM) using cycle-accurate view, architecture view, programmer view, etc. Models at a higher-level (such as programmer view) contain lesser details of the hardware implementation, while at a lower-level (such as cycle-accurate view) contains more details [70].

High Level Synthesis (e.g. [46,95]) allows the designer to write high-level (above RTL) descriptions of a design that can be automatically converted into synthesizable RTL code [123]. High-level code describes the operation on different data types of a hardware design without specifying the required resources and schedule cycle by cycle. Thus, high-level synthesis offers the advantage of automatic handling of the scheduling and synchronization issues of a design, which helps in faster architectural exploration leading to reduced design time and increased productivity.

A typical high-level synthesis process starts by generating the intermediate representations in the form of various flow graphs such as Control Data Flow Graphs (CDFGs) [112], Hierarchical Task Graphs (HTGs) [73], etc. These representations help in capturing the data dependencies and control flow available in the high-level specifications. Various optimizations and transformations are then performed on such flow graphs before generating the RTL code. High Level Synthesis can be performed using a variety of high-level input specifications expressing the behavioral description in different model of computations such as term rewriting systems [28], communicating sequential process [76], synchronous data flow [96], etc. Moreover, different internal representations are used by different synthesis tools to represent these input high-level specifications before converting them into the RTL code. The input specification for HLS is generally provided as a variant of C/C++ such as SystemC [116], Handle-C [43], etc. Apart from the C/C++ based languages, other languages for which high level synthesis support exists include Bluespec System Verilog [28], Esterel [60], etc.

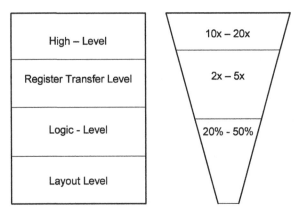

Fig. 1.1 Power savings at various abstraction levels [123]

1.1.2   Low Power Design and High Level Modeling

Each stage of the design flow (shown in Fig. 1.1) represents a different modeling level based on the need of the designer. For example, a high-level stage generally entails the functional description of the design, whereas the RTL includes data flow between registers and logical operation performed on the signals. Gate-level design description include logic-gates (such as NAND, NOR, etc.) and flip flops connected to each other using wires. Layout-level view contains information on how different transistors are connected with each other in various layers of metals, silicon, and other materials. This view is very close to real silicon implementation and contains a lot of design information such as geometric shapes of metal, oxide, semiconductor, etc. Every stage of the design represents the same functionality of the design but expressed in different format, style, etc.

At every stage, designer might have an opportunity to reduce power consumption of the design. However, any small change at the lower level requires more effort to explore possibilities and to validate the design (because of the increased amount of design information), thus increasing the complexity of optimization. Also, at the start of the hardware design process, the designer has more alternatives. At the RTL, micro-architecture of the design is fixed. Fixing micro-architecture implies fixing the data flow between registers, this leaves very few opportunities to optimize the design between/across registers. Based on the constraints on clock-frequency, area, and power, RTL synthesis tools employ some techniques to reduce the power consumption of the design. At the logic and layout-level designers/tools are left with even lesser choices because structure (connection of various transistors) of the design is fixed. "However, no amount of skillful logic design can overcome a poor micro-architecture" [157]. Figure 1.1 presents a documented user experience [123] in reducing power consumption of designs at various stages of the design flow. As shown in the figure most power saving opportunities exist at the highest level of design abstraction.

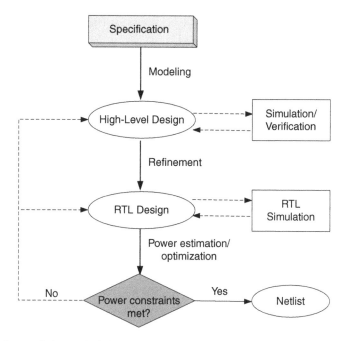

Fig. 1.2 Current design methodology used at the high-level

1.1.3 Current Power Aware Design Methodology at the High Level

Current design methodology used at the high-level is shown in Fig. 1.2 (for simplicity we show power aware design flow, other aspects of a typical design flow are omitted). A high-level design flow starts from a specification. Through rigorous simulations and other verification techniques, a system-level model that conforms to the design intent is constructed. As we know, the RTL design contains a lot more design information than a high-level design; hence, different automated (high-level synthesis) or manual refinement techniques are applied to arrive at the desired RTL implementation (i.e., FSM encoding, micro-architecture variations, etc.). Once the RTL implementation is stable, power estimation of the design is performed and various constraints are evaluated. If the power and other constraints are met, then the netlist is generated. Otherwise, further refinements are applied at the RTL or even at higher levels in order to reduce the power consumption of the design.

1.1.3.1 Problems with the Traditional Methodology

As shown in Fig. 1.2, power estimation is performed at the RTL stage, while refinements to reduce power are applied both at the high-level and the RTL. It may

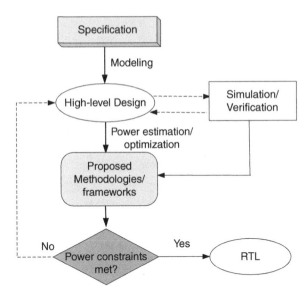

Fig. 1.3 Proposed design flow

be too late to perform power estimation and optimization at the RTL as opportunities to tweak the design to meet the necessary design constraints are limited. High-level synthesis endowed with RTL or lower-level power optimization features (such as clock-gating, operand-isolation, etc.) can instead be utilized in traditional design methodology to exploit the maximum benefits. This will reduce the overhead of repeating the power optimizations at the RTL. This will also help in taking power aware design decisions early in the design flow. There are two main problems in the traditional design methodology:

- Lack of support for accurate power estimation at the high-level
- Lack of support to utilize the RTL power reduction techniques during high-level synthesis

1.1.4   Our Power Aware Design Methodology at the High Level

As solutions to the problems posed above, we propose a design methodology shown in Fig. 1.3. Our proposed design methodology enables power estimation and reduction features at the high-level. Our solutions also include ways in which a high-level model can guide RTL techniques, in case complete migration to a high-level methodology is not possible. We propose the following solutions for both the problems:

- Enable Power Estimation at the High Level:

  – By creating the accurate power model for the hardware design above RTL
  – By utilizing the high-level verification collaterals
  – By utilizing the high-level simulation to guide the RTL power estimation

- Enable Power reduction during High-Level Synthesis:

  – By utilizing the RTL power reduction techniques during HLS such as clock-gating, operand-isolation, sequential clock-gating, etc.
  – By enhancing and utilizing power models to guide the HLS to generate clock-gated RTL

Our proposed methodologies integrate high-level design languages and frameworks. These methodologies are independent of the:

- High-level modeling languages (C, SystemC [116], ESTEREL [23], etc.)
- High-level synthesis tools (Catapult-C [101], Cynthesizer [62], C2R [41], Esterel Studio [60], etc.)
- RTL power analysis tools (PowerTheater [128], Prime-Power [148], etc.)

However, to experiment with our solutions, we have used various frameworks available in academia or industry such as GEZEL [68], C2R [41], Esterel Studio [60], PowerTheater [128], etc.

1.2 Overview of the Proposed Solutions

1. High Level Power Estimation Techniques

   a. *Power Estimation using Statistical Power Models*: A power model is a model that captures the dependence of power consumption of a design block on certain parameters such as switching activity, capacitance, etc. Accuracy of the power model is very much dependent on the model of computation, input/output activity, capacitance, etc. In order to invent a suitable power model for designs, which are modeled as finite state machines with datapath (FSMD). We first analyze its models of computation and propose a power model based on the activity of each state. We capture the relationship between average power and activity of different states of FSMD as a linear regression, to model the power consumption of FSMD based designs [6, 8].

   b. *Power Estimation using High-level Simulation*: Dynamic power is a very important component of total power consumption of a design. It is proportional to the activity/toggles of the design. For power estimation purposes, this activity is generally measured from the simulation of the design. RTL/gate-level simulation of big designs takes a lot of time. These simulation dumps for big designs are typically very large (with the value reaching upto 200

GBytes in some of our experiments). RTL/Lower-level simulation is one of the biggest bottleneck for power estimation because it is very time consuming. To solve this problem, we propose algorithm and methodology to utilize high-level simulation to guide the RTL power estimation techniques. Our methodology helps in significantly reducing the turn around time to obtain the power numbers with reasonable accuracy [11].

c. *Power Estimation using High-level Verification Collaterals*: Verification of a design is very time consuming process. Various verification collaterals such as tests, checker, assertions are used at the early design stages to check the functionality of the design. These collaterals help in finding the appropriate stimulus to check various design states, modes, etc. Measurement of power consumption is very much stimulus dependent. However, there are not many ways in which high-level verification collaterals can be targeted for accurate power estimation tasks. We propose a methodology to apply high-level verification collaterals for accurate power estimation [9, 10].

2. High Level Synthesis endowed with Power-saving Features

a. *Power Reduction using Clock-gating*: Clock-gating is a well known technique for reducing power consumption of the design. This technique is applied to reduce the clock toggles of a register. Clock power is a major component of total power consumption, which makes clock-gating a very important power saving technique. RTL synthesis tools are capable of finding the clock-gating opportunities for a design, if the RTL model conforms to certain modeling guidelines. Since HLS is becoming increasingly important, there is a need to push such lower-level reduction techniques to a higher-level. Here, clock-gating opportunities that can be enabled in a high-level model are investigated. Various granularities of clock-gating at a behavioral level are examined such as function, scope or variable level. An algorithm for clock-gating, to automatically generate clock-gated RTL after HLS is also presented [13].

b. *Power Reduction using Sequential Clock-gating*: There are some techniques such as sequential clock-gating, which require a capability to provide power saving control information across clock boundaries. Verification of optimizations, which are applied across clock-cycles is very difficult. We use model checking to find out the relationship between two registers, and then find out the possibility of sequential clock-gating. Such an approach helps in discovering the relationship between two arbitrary registers. This relationship can then be utilized for sequential clock-gating. Existing solutions are limited to pipe-lined RTL implementations, while our approach is not limited to any particular design-style [12]. Model checking based approach helps in reducing the complexity of verification task, which otherwise is very difficult and limits application of such techniques to particular design style such as pipelined designs. Our approach is further extended to examine such an opportunity to determine sequential optimization opportunities at high-level.

c. *System level simulation guided approach to generate clock-gated RTL*: Clock-gating in the current tools/methodologies is statically done i.e. HLS tool is informed to insert clock-gating for specific part of the design or whole design. Clock-gating may not save dynamic power all the time especially for finer granularities. This savings/wastage is also dependent on the stimulus. In current state-of-the-art methods it is very difficult because power estimation is very time consuming and clock-gating is performed at netlist level, making a decision making task very difficult. Here, a power model capable of checking the efficacy of clock-gating is proposed. This power model indicates how much power savings are possible with clock-gating. Once the indication from the power model is available, an HLS tool is guided to selectively apply clock-gating and generate the new RTL [14].

1.2.1 Application of the Proposed Techniques

In this book, we propose various techniques that can be utilized with HLS or high-level design flows. Here, we briefly discuss the application of the proposed techniques.

1.2.1.1 Early Power Estimation

The approaches discussed in Chaps. 5 and 9 are applicable in scenarios where designers want to measure power consumption of the target hardwares using models that can be utilized at higher abstraction levels than the RTL. Given that RTL or gate-level power estimation is extremely time consuming, such techniques can save time if proven to be accurate enough for crucial design decisions. Chapter 5 presents a technique to characterize power model that is parameterized by the switching activity of each state of FSMD model of the design. Such a power model can then be simulated with high-level models. Chapter 9 shows utilization of directed test vectors and assertions. One could use such vectors or assertions to bring a high level simulation to a target state to measure state specific power consumption.

Since power consumption is very much design and technology dependent, it might be difficult to create power models that characterize power consumption of a target hardware at high-level. To perform power estimation at the RTL, the biggest bottlenecks are simulating, processing and extraction of the activity information. Chapter 8 presents a technique where power estimation is performed using an RTL power estimation tool but the simulation is done at a high-level. The resulting dump is interpolated to create required activity profile. want to utilize RTL or lower level tools to perform power estimation, but the simulation of a design is performed at high-level.

1.2.1.2   Power Reduction from High-Level

We consider clock-gating as a primary candidate for power reduction. We investigate how this can be enabled from high-level. At high-level, a designer is working on a behavioral description of the design. In this case, he/she would need to insert clock-gating from the behavioral description before synthesis. This approach is presented in Chap. 10. Similarly, sequential clock-gating requires optimizations across the register boundaries. Such optimizations may cause a lot of verification problems. In Chap. 11, we propose model-checking based approach utilizing invariants for sequential clock-gating specific optimization. In Chap. 12, we show how system-level simulation can guide HLS to select the appropriate clock-gating candidates among all the registers. This approach is also useful to find out if aggressive application of clock-gating will lead to wastage.

We explore various frameworks in this thesis starting from Cycle-accurate modeling done at as low level as finite state machine with datapath (FSMD) using GEZEL and untimed C that can be represented using C2R for high-level synthesis. We have also explored Esterel for system-level exploration and have shown how various flavors for high-level modeling can be utilized for power exploration purpose.

1.3   Book Organization

1. Chapter 2 discusses related research work done in the domain of high-level power estimation, high-level synthesis, and power aware high-level synthesis.
2. Chapter 3 provides a detailed explanation of various background topics (such as components of average power, GEZEL, ESTEREL, transaction level modeling, C2R based synthesis flow, etc.) relevant to this book.
3. Chapter 4 presents a case study of AES design and possible micro-architectural exploration using HLS from software description of the design.
4. Chapter 5 presents a technique for power estimation using power models of FSMD based design. This chapter details the analysis based on new power model which utilizes datapath activity of each state for power estimation.
5. Chapter 6 presents a methodology for developing power estimation models for FPGAs. Statistical techniques proven to work with Athe SIC designs is studied in the context of FPGAs. The power model weighs in both the design details, in the form of toggle counts and the FPGA resource consumption by the designs. Also, the conventional techniques work on developing power models for individual designs. This method, explores the possibility of obtaining generic power model for multiple designs.
6. Chapter 7 demonstrates case studies for using high level synthesis to perform power-performance trade-offs in Hardware–Software codesign. The inherent flexibility available in higher levels of design abstraction is harnessed to move components between hardware and software platforms to obtain quick estimates.

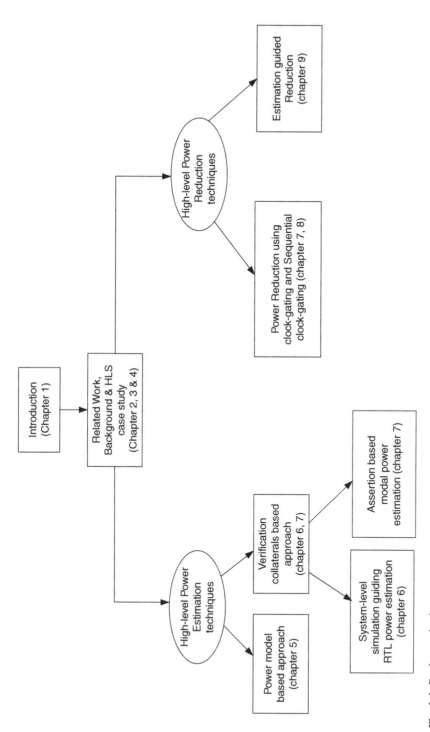

Fig. 1.4 Book organization

7. Chapter 8 presents a technique to obtain early power estimation using vectorless RTL power estimation technique guided by high-level simulation.
8. Chapter 9 presents a technique which utilizes verification collaterals for accurate power estimation. This chapter presents a case study which helps in understanding verification collaterals needed for accurate power estimation. These collaterals include directed testbenches, assertions, which helps in obtaining state-specific power numbers.
9. Chapter 10 presents power reduction using clock-gating at the high-level. This chapter provides detail on how clock-gating can be enabled in high-level specification. This chapter also presents an algorithm for clock-gating used in high-level synthesis.
10. Chapter 11 provides an overview of sequential clock-gating based power optimization. We show how model-checking can be utilized to find the opportunities for sequential clock-gating from the behavioral description and further utilizing the indications after checking few properties of the design to generate sequentially clock-gated RTL.
11. Chapter 12 presents an approach where power model guides power reduction. We present a power reduction model for clock-gating based power saving technique. This power-model is embedded in cycle-accurate SystemC view of the behavioral model. This model gives an indication on the registers that can save power and further guides the HLS to generate the clock-gated RTL. This chapter presents the complete methodology and theory for creating the power reduction model.
12. Chapter 13 presents conclusions and the possible extension for the approaches presented in this thesis.

Figure 1.4 shows the organization in a pictorial form along with the separation of the work that is completed and the future work. The organization shown in Fig. 1.4 presents the content dependencies to guide the reader in understanding the material that is of interest to them.

Chapter 2
Related Work

In this chapter, we briefly discuss the existing techniques and methodologies in the area of power estimation, high level synthesis, and power reduction.

2.1 High Level Power Estimation

Accurate power estimation at high level is very important for any successful design methodology. This area has been extensively researched. In this section, we capture some research advancements relevant to this thesis, which include spread sheet based approach, power model, and macro-model based approach, commercial tools available for RTL and gate-level power estimation, etc.

2.1.1 Spreadsheet Based Approaches

Spreadsheets are very useful in the early stage of design process, when initial planning is going on and a lot of important decisions are being taken [158]. One of the biggest advantages of spreadsheet based analysis is that the user does not really need to learn any complex/sophisticated tool for taking design decisions. One of the basic application of spreadsheet is area estimation. Designers generally have a fair idea of the building blocks for a big design. He/she can easily get an estimate on area by using data sheets from intellectual property (IP) provider, library cell estimates, etc. Spreadsheet provides a capability to capture such information, which can be utilized for quick area estimation. Similarly, some decisions to control power can also be taken using spreadsheet based approach. Power budgeting approaches using spreadsheets are very helpful for printed circuit board (PCB), power supplies, voltage regulators, heat sink, and cooling systems.

S. Ahuja et al., *Low Power Design with High-Level Power Estimation and Power-Aware Synthesis*, DOI 10.1007/978-1-4614-0872-7_2,
© Springer Science+Business Media, LLC 2012

Spreadsheet tools vary from utilizing excel sheets, word processors to Unified Modeling Language (UML) [153], etc. In industry, spreadsheet is being advocated by Field Programmable Gate Array (FPGA) vendors such as Xilinx [163], Altera [121]. Power analysis needs to be done very efficiently especially for FPGAs, where basic building blocks are fixed (for example, fixed size lookup tables and switch matrices). These power estimation tools provide current (I) and power (P) estimation for the different family of FPGAs. Power number associated to each block in these spreadsheets are very much dependent on architectural features such as the clock network, memory blocks, Digital Signal Processing (DSP) blocks. User can enter operating frequency, toggle rates, and other parameters in the early power estimator spreadsheet to estimate the design's power consumption. Spreadsheet based approach is useful for project planning but may not be able to provide accurate guidance for block-level hardware power estimation and reduction. This motivates a need to provide a power model, which can perform accurate yet efficient power analysis early in the design flow. The next subsection provides an overview of some of the model based approaches for power estimation purposes.

2.1.2 Power Estimation Approaches Utilizing Power Models

One of the first works in the area of model based power estimation was proposed by Tiwari et al. [151] for the architecture level power estimation of microprocessors. They provide a method to estimate power/energy number of a given program on a given processor. The way they have approached the problem was interesting, they have used a hypothesis for their work. This hypothesis states that "By measuring the current drawn by the processor as it repeatedly executes certain instructions or short instruction sequences, it is possible to obtain most of the information that is needed to evaluate the power cost of a program for that processor" [151].

Power consumption of a microprocessor can be represented as $P = V_{CC} \times I$, where V_{CC} is operating voltage and I is the current drawn by the microprocessor. In their approach, they measured the current drawn by the processor and then utilize it for power measurement purpose. They assumed that during the computation, the operating voltage for the processor will not change. For average power estimation purpose they had first estimated energy over different cycles and then averaged it. To conclude with they have proposed a way in which one can gauge the impact on average current from the execution of an instruction. Their approach also proposed a way to measure inter-instruction effects on current values. To calculate the average power consumption of a processor while executing a software program, they utilized these values.

Power modeling for the instructions of Very Large Instruction Word (VLIW) processors is discussed in [35]. Approach discussed by Tiwari et al. in [151] might not be applicable to the processors where number of instructions are reaching to a few thousand. The valid sequences for such processors can reach thousands. Characterization of power instruction based power model can be very time consuming and may take several months. Tiwari et al. in [151] propose a technique for VLIW processors for reducing the complexity of power models by using the clustering approach. They try to make a cluster of instructions having energy number in the same range (individual as well as sequence of instructions). They propose an approach to reduce the complexity of characterization from exponential to quadratic. There are many other approaches that have been discussed in the literature for doing power modeling of processor instructions. More details on power estimation utilizing instruction level power models and its variants are available in [21, 126, 150].

Most of the approaches for power models are proposed for CPU or micro architecture of processors. In an ASIC design flow where mainly design stage starts at RTL and mostly design is represented as a Finite State Machine with Datapath (FSMD), similar approach may not be useful. In the literature, this is often termed as macro-model based approach. In the Sect. 2.1.3, we present an overview of existing macro-model based power estimation approaches. We have used quite a few concepts from such approaches in this thesis. A quick read of the Sect. 2.1.3 will help the reader in understanding the approach discussed in the Chap. 5.

2.1.3 Power Macro-Model Based Approach

Power macro-models [74] generally consist of n-dimensional table to estimate the power consumed in a circuit for a given statistics, where n represents different variables/components capturing the relationship of power and dependent variables such as input probability, transition density, etc. [74], present an automation methodology, in which, such a table is automatically generated. Three variants in their model are average input signal probability, average input transition density, and average output zero-delay transition density. Power macro-model is a function of these variants, as shown in (2.1).

$$P_z = f(P_{in}, D_{in}, D_{out}) \qquad (2.1)$$

where P_z represents the entry for the average power corresponding to the average input signal probability P_{in}, average input transition density D_{in}, and average output zero delay transition density D_{out}.

Signal probability P_i at input node x_i is defined as the average fraction of the clock cycle in which final value of x_i is high. Similarly, transition density D_i at an input node is defined as the average fraction of the cycles in which the node makes a logic transition (final and initial value should not be the same for a logical transition). Equation (2.2) represents the relationship between P_i and D_i.

$$D_i/2 \leq P_i \leq 1 - D_i/2 \qquad (2.2)$$

Using the relationship discussed in (2.2) and (2.1), characterization of models is being done and the accuracy of these macro-models is evaluated. For characterization purposes they assume that input nodes are output of latches or flip-flops and make at most one transition per clock cycle. Also, sequential design is single clock system and clock skew is ignored in their analysis hence all the inputs switch simultaneously.

Bogliolo et al. [34] have proposed a methodology for creating power macro-models based on linear regressions but their flow is specific to the structural RTL macros and power estimation is done at the gate-level. Their analysis is restricted to structural RTL representation whose leaf components are combinational logic blocks. This approach is based on (a) offline characterization in which they compute the power of the RTL macro based on certain tests and (b) online characterization, in which they do it adaptively for error minimization. Their approach utilizes all the inputs, outputs, transition functions of inputs and outputs on the successive cycles, and then they interpolate the relationship with energy consumption.

Potlapally et al. [119] present a technique in which they do cycle-accurate power macro modeling of the RTL component. This technique is based on the fact that RTL components exhibits different "power behavior" for different input scenarios. They create power macro model for each of these behaviors also known as power modes. Their framework chooses the appropriate power mode from the input trace in each cycle and then apply power macro-modeling technique discussed by Bogliolo et al. to get an estimate on power numbers. The technique discussed in [34] is limited to the typical average power estimation scenario while the technique in [119] covers non-trivial scenarios as well.

Negri et al. [113] have proposed a power simulation framework of communication protocols (Bluetooth and 802.11) using StateC. StateC is used to model the hierarchical state machines. Their flow is mainly targeted for simulator generation in SystemC. This flow is good for power exploration of protocol modeling but not presented on ASIC/FPGA design flow such as ours. They have mainly targeted wireless protocols in which relevant contribution to the power consumption of a node is due to the communication and not due to the datapath (computation) activity. Their learning phase requires execution of the real chip and can not easily be integrated to any ASIC/FPGA design flow. Also for practical purposes, it is difficult to create power model of the partial design or smaller part of the whole chip because current measured includes a lot of contribution from the other parts of the chip and isolation of the desired unit for power model purposes requires quite a lot of effort.

In [164], authors attempt to lift power estimation to higher levels than the RTL, and their choice for high-level modeling was Cycle-Accurate Functional Description (CAFD) of the design. They create virtual components for each design block and attach them to the CAFD model of the design block, and compute the power consumption dynamically as the CAFD is simulated. Since this additional overhead to the CAFD simulation causes inefficiency, they also allow periodic turning off of some of the virtual components during some cycles of the simulation. During those cycles, they estimate power based on the history of the power consumption for the turned off components. So even though, the abstraction at which they estimate power is cycle accurate modeling level as ours, their power estimation is not based on regression based technique, and the simulation of CAFD is slower due to the overhead of virtual components.

Caldari et al. presented a relative-power analysis methodology [40] for system-level models of the Advanced Micro-controller Bus Architecture (AMBA) and Advanced High-performance Bus (AHB) from ARM. It relies on creating macro-models from the knowledge of the possible implementations. Similarly, Bansal et al. presented a framework in [20], which uses the power-models of the components available at the system-level simulation stage by observing them at run time. It selects the most suitable power-model for each component by making efficiency and accuracy trade-offs. In [93], the presented framework employs co-simulation techniques for power estimation with the capability of performing accuracy and efficiency trade-offs. They utilize multiple power estimation engines concurrently and have proposed several speed-up techniques to eliminate the overhead of co-simulation.

SEAS [22] presents a system and framework for analyzing SoCs in early design stage. Power Analysis in this system works at a granularity of a processor cores, where pre-characterized data for power is utilized based on the power state of the design. Power states of the cores are high level states based on the workload such as active, idle, sleep states in today's processors. By utilizing these high level states of the SoC an early power estimation can be performed which is more efficient and accurate than the traditional spread sheet based approach.

Shin et al. have proposed a methodology [131] for power estimation of operation modes but their analysis is done at logic-level and proposes a way to create power models based on the switching frequencies. Our approach is different from their approach as we are targeting the modal designs for power estimation at system-level keeping good accuracy in mind. [30–32,75,100,114] utilize the similar approach for power estimation purposes and provide various accuracy and efficiency trade-offs based on the quality of inputs and power modeling. Power estimation accuracy can be significantly increased but generally it impacts the efficiency of power estimation procedure.

2.1.4 Summary of Power Estimation Research

Techniques discussed above show that power modeling of a hardware block can be very complex and application dependent. As we increase the levels of abstraction, power estimation of a hardware block becomes difficult. Various techniques at different abstraction levels exist to obtain the power consumption starting from spread-sheet, power model to macro-model based power estimation. The most popular approaches in industry are mainly power model based approach or by performing power estimation at RTL/gate-level description of a hardware design. While, at the lower level of abstraction, commercial tools provide good accuracy with respect to silicon, but as we go to higher levels of abstraction accuracy of the power estimation methodologies reduces.

In an ASIC design flow, most of the mentioned power-estimation solutions either rely on the creation of power macro-models from lower level descriptions of the designs or on the availability of power consumption information for different simulation scenarios. The methodologies we propose in this thesis are not dependent on any such infrastructure, thus making it suitable for easier and accurate power analysis at early stages of an ASIC design flow. Previous approaches in this area are focused on creating macro-models for the design based on input activity and relationship with the corresponding power. Such input and output patterns help in modeling power of the design but design is considered as *blackbox* in most of the cases. However, approaches used in this thesis provides explicit visibility on the toggles and corresponding control state and datapath, and can be viewed as a gray box approach.

In this thesis, we also present approaches, where high level fast simulation is guiding lower level tools. Most of the power estimation tools rely on simulation and simulation dump processing, which takes majority of power analysis time. We investigated techniques that work with high-level simulation to provide the maximum benefits. We experimentally show very high accuracy can be achieved.

2.2 High Level Synthesis

Transistor level designs were prevalent in 1970s, gate-level in 1980s, and RTL in 1990s, after RTL there is no consensus. RTL based designs have increased designer's productivity as newer chips are utilizing more and more transistors in the same design. Managing such a design complexity requires a lot of effort on design and verification side. Engineering budget to include more and more engineers at lower level along with the increasing cost of mask is another compelling reason to automate the process from the earliest possible design stage. This brings a need for Electronic System Level (ESL) based design approach, where design, verification, and prototyping is envisioned at a level of abstraction where designer's concern is to take care of the best possible algorithm to complete the task while other tasks such as creation of hardware, functional verification, validation, etc., will be seamlessly integrated to the design flow. One of the most important tasks for achieving this goal is to be able to produce hardware from C/C++ based specification or the specifications which are above register transfer level (RTL) abstraction of the design.

High-level synthesis is an emerging area. There are lot of tools and methodologies exist in the industry and academia. Most of these tools are in infancy stage. In hardware design industry, adoption of these tools has recently started. High-level synthesis of hardware designs has been shown to be possible from a variety of high-level languages. These languages are used to specify the behavior of a design at a level of abstraction above RTL. Hardware descriptions written using such HDLs are passed as inputs to high-level synthesis tools to generate the RTL code. There are various ways high-level synthesis can be enabled but most of the contributions are in the area of C/C++ based synthesis.

The other motivations to designing at high-level with respect to Verilog, VHDL for hardware design are

- Most of the HDLs are foreign to algorithm developers (they generally prefer C/C++).
- Increasing level of abstraction using HDLs is very difficult. As most of the times, without the availability of FSM/FSMD, a good RTL cannot be developed for the design.
- In practice, HDL based design flow is time consuming. Generally, the golden model (from which functional correctness is verified for the lower level models) comes as an executable, which is C/C++ based implementation of the design. Hence an automatic path from the C/C++ implementation would first reduce the time. Secondly, once flow is stable, it will reduce a lot of verification effort, because generated RTL needs to be validated every time with respect to functional specification for any small manual change done at RTL.

2.2.1   High Level Synthesis from C/C++ Specifications

Here, we briefly touch upon the existing high-level synthesis tools and methodologies used in industry and academia generating hardware from C/C++ specifications. In this section, we cover in detail the methodology presented by CatapultC and PICO tool. We also provide a brief overview of some other tools.

2.2.1.1   Catapult

In Catapult synthesis [101], flow various manual methods are automated to reduce the time to produce the hardware. The flow is centered around Catapult HLS tool, where micro-architectural selection is done based on the constraints (provided by the designer/user). Tool creates the RTL based on these constraints. Other important point to note is that one can specify the target technology used, clock-period or clock frequency as constraints. Such information helps the scheduler of the HLS to schedule and allocate resources based on the constraints. Below, we briefly discuss various important points of Catapult based design methodology.

- *Verification of generated RTL against original C code*: Functional verification of the generated RTL is one of the most important stage of the design flow. Here, RTL is wrapped around a SystemC transactor, where it is called as foreign module. By doing this, original C++ testbenches can be compiled with this SystemC top module instantiating generated RTL module and finally comparator is used to compare the outputs. The wrapper code along with the makefiles is auto-generated to complete the verification flow.

- *Synthesis Constraints for Catapult Flow*: There are two types of constraints that can be provided in the Catapult flow. First set of constraints are related to target technology, clock-frequency, etc. Second set of constraints are utilized to control the architecture. These constraints can be inserted using GUI or using directives. These directives facilitate loop-unrolling, loop pipelining, hardware interface synthesis, etc. These constraints are not encoded in the source code; hence appropriate micro-architectures are created during synthesis stage.
- *C++ and optimization support*: Catapult supports the pure C++ based specification as input. Most of the C++ constructs are supported by the tool except the code, which requires dynamic memory allocation/de-allocation such as use of malloc, free, new, delete. In other words, code should be statically deterministic so that all the properties, memory allocation can be performed during compile time. Catapult supports pointer synthesis, classes, and templates, bit-accurate data-types, etc. On the optimization side Catapult C supports loop pipelining, loop-merging, loop unrolling, technology driven resource allocation and scheduling, etc.

2.2.1.2 Design Flow for PICO Tool

Program In Chip Out (PICO) tool's [147] approach is mainly targeting SoC platforms. Before going to the discussion of tool it will be helpful to see the characterization of different IPs in the context of this tool and their approach. These IPs are mainly characterized in four categories as discussed below:

- *Star IPs*: Star IPs are CPUs/DSP blocks and generally these blocks are fixed for many generation of SoCs. Most of the times these IPs are manually created and their instruction level characteristics are well defined. During the development of SoC various models of such IPs are utilized at different granularities, which includes Instruction level simulation model, RTL/gate-level model, etc. These IPs are not altered for design changes.
- *Complex Application Engines*: Especially for embedded systems, complex application IPs such as video-codec, wireless modem, etc., are differentiating factors for the end product. These codecs or modems generally change as new standards are introduced. Functionality of these complex IPs require small to significant changes for a new SoC development and in many cases direct utilization of these IPs may not be advisable.
- *Connectivity and Control IPs*: This set of IPs include connectivity and control IPs, examples of such IPs include DMA, USB, etc. These IPs are generally utilized in communication and it can be considered as system-level glue. Their functionality is not needed and requires very minimal tailoring.
- *Memory*: Memory is generally the biggest contributor to silicon area and generally at system-level their functionality is not defined. Its functionality generally does not differentiate the end product. Memory models are compiled and built bottom-up.

All the IPs discussed above are essential for SoC development but complex application engine generally require the bulk of effort for design and verification purposes.

2.2.1.3 Programming Model of PICO

PICO can accept the sequential C specification and tries to extract the parallelism from the sequential C code. In that sense, their approach is very much specific to particular application domain such as signal processing applications. Because in such domains a lot of parallelism is available while application is processed by the hardware. Such a programming model is useful where a part of function has no dependency between the different tasks. If one task works on a block of data and other works on another block of data, then a lot of parallelism can be extracted in pipelined manner. The execution model of PICO is based on Kahn Process Network (KPN), where set of sequential processes communicate via streams with block-on-read and unbounded buffering. Here, processes are the hardware blocks, which communicate with each other through streams. The restriction on unbounded buffering is a big restriction, which is basically solved by imposing additional constraints on the execution model.

GAUT: GAUT [66] is an academic high-level synthesis tool based on C as design language but applicable for digital signal processing applications. GAUT starts from bit-accurate specification written in C/C++ and extracts the possible parallelism before going through the conventional stages such as binding, allocation, and scheduling tasks. GAUT has mandatory synthesis constraints, which are throughput and clock-period. GAUT utilizes the algorithmic C libraries (from mentor graphics) for bit-accurate integer and fixed-point data-types using ac_int and ac_fixed data types. HLS in GAUT utilizes the DFG generated from gcc and library characterization for particular target library for area, speed, etc., to generate the RTL VHDL or cycle accurate SystemC. The synthesis process is not completely technology independent but can be useful for virtual platform development for micro-architectural analysis.

SPARK: SPARK [73] presents a high-level design methodology based on the high-level synthesis tool, which takes Behavioral C as input design language and capable of generating RTL VHDL. Main contributions of the methodology based on SPARK are (a) inclusion of code motion and code transformation techniques at compiler level to include maximum parallelism for high-level synthesis and (b) proposal of high-level synthesis starting with behavioral C input. The approach presented in SPARK helps designer in understanding how and by what amount the quality of results can get affected by the language level transformations such as loop unrolling, loop-invariant code motion, etc., on generated circuit from HLS. Also, the approach presented in SPARK based methodology suggests that no single code transformation approach is universally applicable but such techniques with heuristics applied for particular application domain leads to better quality or results. The transformations and techniques applied in this methodology include exploitation of instruction-level parallelism, such as speculative code motions, percolation scheduling and trailblazing, etc.

C2R: C2R [41] is also a tool, which takes ANSI C based description as input and capable of producing synthesizable RTL Verilog as output. C2R benefits the designer in two folds: (a) facilitate algorithmic developer to remain in C/C++ native execution environment and (b) provide directives to include concurrency, implementing interfaces using C functions, exploiting micro-architectural parallelism, concurrency inside a function/process, facilities to exploit pipelining behavior, etc. Functional verification of the design is performed on the generated RTL output by comparing its output against the output generated from the pure ANSI C based implementation. Other advantage tool provides is instantiation of complex memories in behavioral style in pure ANSI C environment. Such an approach helps SoC development where parts of the designs are IPs coming from other vendors and some can be developed in-house quickly. Finally, Cebatech [41] provides a methodology to designer to exploit the multi-threaded C code using thread and convert it into Verilog RTL.

2.2.2 High Level Synthesis from Behavioral Specifications

Bluespec [27] provides high-level synthesis solution based on Bluespec system verilog (BSV), which is based on atomic transactions. The compiler is capable of generating synthesizable HDL from BSV. The reason for using BSV instead of C/C++ is because they are not considered appropriate for certain control dominated applications. Such applications include processors, caches, interconnects, bridges, etc. Secondly, atomic transaction based model of computation suits well to some of the application mentioned above, which otherwise may be very complex if implemented using conventional HDL based approach. For verification BSV based methodology can either utilize the simulation based approach or formal verification based approach, which can be applied to the atomic model of computation, such verification methods are based on term rewriting systems, SPIN model checker, etc. However, designers need to learn a new language to develop the hardware using Bluspec.

Cynthesizer [61] provides the high-level synthesis capability around SystemC transaction level models. The main contributions of this tools comes in the synthesis of transaction level models. On the verification/simulation side, since the input language is SystemC, for the synthesizable subset of SystemC designer gets the flexibility to utilize the C++ execution environment. SystemC has support for processes using sc_thread, sc_method, sc_cthread, etc. Also, SystemC helps designer to write efficient interfaces for communication between computation blocks. Such a high-level synthesis approach is very helpful in system design where designer wants to utilize IPs/hardware blocks and utilize various communication interfaces to develop SoC. C++ execution helps in quick verification and generation of interface and hardware blocks to RTL quickly, which helps in improving the design cycle efficiency.

Apart from above mentioned contributions, various techniques for hardware synthesis from many C-like languages have been proposed [53,57]. Most techniques either use a subset of C or propose extensions to C in order to synthesize the high-level code to RTL. Cones [145] synthesis system from AT&T Bell Laboratories allows the designer to describe the circuit behavior during each clock cycle of a sequential logic. It takes a restricted subset of C (with no unbounded loops or pointers) as input for hardware synthesis. Similarly, HardwareC [91] is a C-like language having a cycle-based semantics with support for hardware structure and hierarchy. It is used as an input to Olympus synthesis system [54]. Celoxica's Handel-C [43] is a superset of ANSI-C with support for parallel statements and rendezvous communication. SpecC [55, 65], which is a system-level design language based on ANSI-C, has an explicit support for hierarchy, concurrency, state transitions, timing, and exception handling. SpecC's synthesizable subset is used to generate hardware designs. Other tools and languages, which are mainly used in high-level modeling and for which synthesis support exists include Celoxica's Agility Compiler [42], NEC's Cyber [156]. ESTEREL [25] is a synchronous language, allows the description of a design at high-level, which can then be synthesized using Esterel Studio [60]. ESTEREL supports implicit state machines and provides constructs for concurrent composition, preemption, and exceptions [24,56].

There are various approaches proposed in software synthesis area by Jose et al. The proposed techniques in this thesis or other related works can be adopted for such synthesis frameworks as well. Synchronous languages such as Esterel, Lustre, etc. can also be used to generate embedded software [89]. Synthesis tools such as Esterel Studio and SCADE are used to generate control software in safety critical applications like avionics, power plants, and so on. An alternate programming model called polychrony, which deals with multi-rate signals whose event occurrences are independent of the software execution rate is also used for generating embedded software. The programming language SIGNAL [71] based on relational models or the actor based network MRICDF [85, 86] can be used to capture polychronous specifications. These programming formalisms are associated with their software synthesis tools Polyrchrony [59] and EmCodeSyn [83, 84], respectively. To obtain efficient embedded software implementations on distributed environment, the generation of multi-threaded code [82,87] and the communication between distributed synchronous systems are also being researched [88].

2.2.3   Summary of HLS Research

A lot of initial work for HLS started in academia but unfortunately the quality of the hardware generated was not very good. However, in the recent past more and more design wins are shown in industry using HLS tools. HLS tools are maturing day-by-day. Initial research in this area was about finding good model of computation and language to represent the hardware. A lot of tools have shown success in producing hardware from the behavioral description in various different modeling languages. At the time of writing this thesis, I can think of atleast two viewpoints for different languages to represent the design at high-level:

- Use C/C++ as design language because most of the specifications in practice come in C/C++. A lot of companies have kept this viewpoint in mind and have introduced HLS tools, which require minimal code restructuring to generate synthesizable RTL. Such examples include tools like CatapultC, PICO, C2R etc.
- Use the behavioral description amenable to verification or particular model of computation. This helps in creating hardware, which can be easily verified in the later design stages. Also, it greatly helps in resolving issues, which exist in the behavioral description in C/C++ itself but it introduces an intermediate step of creating another description in different language. Examples include Bluespec compiler for Bluespec system verilog, Forte Cynthesizer for SystemC, Esterel Studio for Esterel language, etc.

In this thesis, we have used tools from both the viewpoints. We have used Esterel Studio and C2R. We have shown power measurement and reduction approaches for these tools. Most of the proposed approaches in this thesis, can be seamlessly integrated with tools accepting C/C++ or some other behavioral description as the modeling language.

2.3 Power Reduction at the RTL and High Level

Substantial research has been done in the area of power reduction at the RTL and higher-levels of abstractions. Clock-gating is implemented at the RTL by tools such as PowerTheater [129], Power compiler [120], etc. PowerTheater suggests opportunities to clock-gate registers for which there is no multiplexer in the feedback path. They find the conditions under which clock is required when and only when there is a change in data input to a register. Power compiler recommends particular RTL coding styles to enable clock-gating during synthesis.

Such local optimization opportunities are often found dynamically through simulation, and they do not exist in HLS. This is because (1) in HLS enable signals for clock-gating purpose can be generated at the compile time without any dynamic analysis or symbolic simulation and (2) global enable for the registers is easily visible at the high-level because at lower level conditions under which a register is updated can not be determined, such as conditional execution of blocks using "if–then–else," "while," etc. Reference [58] provides an overview of clock-gating technique and how tools like power compiler use it. They also show the advantages of clock-gating for register banks and how it helps in the place and route stage of the design. Many of these concepts are used in our work as well albeit at a higher level. However, this work requires RTL simulation, generation of VCD, and subsequent analysis, making the entire process extremely time consuming. At the higher level an effort is required to make the entire process few orders of magnitude faster, while power savings should be better or equal atleast.

High-level synthesis from C-like HDLs commonly use Control Data Flow Graphs (CDFGs) as intermediate representations during the synthesis process, and consequently, most research in the area of low-power high-level synthesis is targeted towards the CDFG-based synthesis. In the past, various power optimization techniques targeting the power reduction during synthesis have been proposed. High-level synthesis for C-like HDLs include stages such as scheduling, allocation and binding. Various techniques are proposed for different stages to affect the power consumption of the design once the RTL is created from HLS. Scheduling of various operations of a design can be exploited for generating power-efficient designs. The problem of resource-constrained scheduling for low-power has been addressed in [92, 134]. These approaches use CDFGs to first determine the mobility of various operations based on the ASAP and ALAP schedules. Using the computed mobilities and other relevant factors, priorities are assigned to various operations. Based on the assigned priorities, various operations of the design are then scheduled in each clock cycle such that the power consumption of the design is reduced.

During the allocation phase of a high-level synthesis process, functional units and registers are allocated to the design, whereas in the binding phase, operations and variables of a design are mapped to the allocated functional units and registers respectively. References [48, 95, 111, 122] present techniques targeting low-power reduction during allocation and binding phases. Reference [122] presents an allocation method for low-power high-level synthesis, which selects a sequence of operations for various functional units such that the overall transition activity of the design is reduced. Reference [111] presents an algorithm targeting the minimization of the average power of a circuit during the binding phase of high-level synthesis process using game-theoretic techniques. In that work, binding of the operations to the functional units is done such that the total power consumption is minimized. Reference [95] presents an efficient register-sharing and dynamic register-binding strategy targeting power reduction in data-intensive hardware designs. The experiments demonstrate that for a small overhead in the number of registers, it is possible to significantly reduce or eliminate spurious computations in a design. The proposed strategy handles this problem by performing register duplication or an inter-iteration variable assignment swap during the high-level synthesis process.

Reference [58] provides an overview of clock-gating technique and how tools like power compiler use it. Many of these concepts are used in our work as well albeit at a higher level. Reference [15] discusses how observability don't care (ODC) conditions can be exploited for reducing the switch activity for redundant computations. Such an approach provides more advantage over the one provided by the commercial tools. Most of the RTL based approach shows that inability of designer to look into the control flow of the design at high-level leads to such opportunities. Also, many times, various different blocks of the design are created by different RTL designers because of which sometimes global clock-gating or enabling conditions are missed by the RTL clock-gating tools. Reference [63] utilize input stability conditions to find more aggressive power saving opportunities than ODC based approach. Cong et al. [49] show the application of ODC and STC for HLS tool. In all the analyses, simulation based selection is not performed. Our approach can be considered as a co-operative solution because we provide a guidance on top of the selections made by such techniques.

Munch et al. discuss the opportunities to reduce power at the RTL using operand isolation based technique to reduce the dynamic power at the RTL [110]. Operand isolation is a technique, which helps in reducing the redundant activities around datapath unit and is considered as a complementary technique to clock-gating. Because clock-gating does not help in controlling the datapath activity, it just controls the clock toggles of registers.

Reference [3] describes a system called CoDel, where they apply clock-gating at high-level to DSP benchmarks for power reduction. The reported power numbers show some clock-gating opportunities, which Power Compiler misses but their tool can find. They utilize state transition information and set of reads and writes for each state to obtain the clock-gating logic. Although their approach helps in reducing power consumption, reported numbers show almost 20% area penalty and 15% timing penalty as compared to Power Compiler. timing and area penalty w.r.t. clock-gated design because we do not clock-gate every register. In the approach presented in chapter 10, 12, we analyze the CDFG during HLS to find out the common enable signals for sets of registers so that instead of individually clock gating registers, banks of registers can be clock gated by the same logic. These optimizations are included in our proposed algorithms. We show using our approach, 2-3 orders magnitude faster analysis can be performed.

Reference [102] also provides an approach to enable clock-gating in the HLS flow but their approach also lacks a simulation driven realistic power reduction feature. Their approach require RTL synthesis to insert the gating logic while we insert the necessary logic into the source code before generating the RTL. Singh et al. in [137] present algorithm for clock-gating and operand isolation to reduce the dynamic power consumption of the design at high level in Bluespec [28]. They propose a technique to automatically synthesize power optimized RTL using Bluespec compiler. Singh et al. present quite a few approaches for power reduction for concurrent action oriented synthesis tools such as Bluespec in [26, 78, 135–143]. They have presented approaches for dynamic power reduction using operand isolation and clock-gating. They discuss the complexity of power reduction using clock-gating and rescheduling of rules. They also present an approach in which rescheduling for rules is performed to reduce the peak power consumption of the design. Further, this may impact the functionality of the design for which they present a formal verification approach using a model checker SPIN. They check the equivalence between the design before and after performing the optimization.

Clock gating and operand isolation are two techniques to reduce the power consumption in state-of-the art hardware designs. Both approaches basically follow a two-step procedure: first, they statically analyze a hardware circuit to determine irrelevant computations. Second, all parts which are responsible for these computations are replaced by others that consume less power in the average case, either by gating clocks or by isolating operands. Jens et al. [36] defines the theoretical basis for adoption of these approaches in their entirety. They show how irrelevant computation can be eliminated using their approach. They present passiveness conditions for each signal x, which indicate that the value currently carried by x does not contribute to the final result of the system. After showing how their theory can be generally used in the context of clock gating and operand isolation a classification of many state-of-the-art approaches is performed and shown that most of the approaches in the literature are conservative approximations of their general setting.

Reference [48] targets low-power design for FPGA circuits. It presents a simulated annealing engine that simultaneously performs scheduling, resource selection, functional unit binding, register binding and datapath generation in order to reduce power. Reference [48] also proposes a MUX optimization algorithm based on weighted bipartite-matching to further reduce the power consumption of the design. Power management refers to techniques that involve shutting-down parts of a hardware design that are not being used for power savings. This can be done by disabling the loading of a subset of registers based on some logic [47, 50, 108, 110]. Operand Isolation (also known as *signal gating*) avoids unnecessary computations in a design by gating its signals in order to block the propagation of switching activity through the circuit. Reference [47] discusses automation of operand isolation during Architecture Description Languages (ADL) based RTL generation of embedded processors. In [110], a model is described to estimate power savings that can be obtained by isolation of selected modules at RTL.

Reference [50] defines power-island as a cluster of logic whose power can be controlled independent from the rest of the circuit, and hence can be completely powered down when all of the logic contained within it is idling. Reference [50] proposes technique that eliminates spurious switching activity and the leakage power dissipation in a circuit through the use of power islands. The technique first schedules the design in order to identify the minimal number of resources needed under the given latency constraints. After scheduling, the functional unit binding is done based on the life cycle analysis of all the operations. The scheduling and binding steps are followed by a partitioning phase which uses a placement algorithm that performs partitioning such that the components with maximally overlapping lifetimes are clustered together, and assigned to the same power-island. After the partitioning phase, register-binding is performed such that both the total and the average active cycles of registers are minimized.

Recently power reduction techniques have also been used for security purpose. With increased outsourcing, confirming the genuineness of third party manufactured ICs has emerged as a major challenge. Researchers have effectively used various side-channel analysis techniques to fingerprint ICs viz. power, timing, EM-radiation. In [16, 17] authors have used circuit partitioning techniques to selectively exaggerate power consumption in targeted portions while reducing the overall power of the chip. In [18] a sustained vector scheme is employed to ensure that no transitions are initiated from the PIs so that the overall transitions occurring (that translates to overall dynamic power) in the circuit can be minimized. In [19] authors have used voltage inversion technique to change the functionality of the circuit so that malicious insertions are exposed off. This is augmented by a power profiling technique which is proven to be very effective in light of the changed functionality.

Some other low-power high-level synthesis works include [45, 90, 152]. Reference [90] presents a high-level synthesis system for targeting reduction of power consumption in control-flow intensive as well as data-dominated circuits. The system uses an iterative improvement algorithm to take into account the interaction among the different synthesis tasks for power savings. Reference [45] presents a high-level synthesis system for minimizing power consumption in application specific datapath intensive CMOS circuits using a variety of architectural and computational transformations. The proposed approach finds computational structures that result in the lowest power consumption for a specified throughput given a high-level algorithmic specification. Reference [152] proposes a thread partitioning algorithm for low-power high-level synthesis systems. The algorithm divides parallel behaving circuit blocks (threads) of a design into subparts (sub-threads) such that gated-clocks can be applied to each part for power savings. As proposed in [106, 107, 124, 132, 133], high-level synthesis of hardware designs can also be used to target the peak power reduction of the generated designs.

Lakshminarayana et al. [94] present a methodology to quickly explore various hardware configurations and also to obtain relatively accurate design matrices for each configuration. They use C2R high level synthesis tool to quickly generate RTL description of the hardware. They show how HLS languages give a greater degree of freedom to do selective optimization by means of inserting the so-called "directives" into the behavioral code (also known as restructuring). They present case studies to develop or modify behavioral IP descriptions and use standard FPGA boards to profile the IP in very short time. The difference in the measured and actual IP design matrices are not significant as one is more concerned with relative difference among various configurations. A variety of compute-intense benchmarks like AES is used to demonstrate how platform specific optimizations as well as higher level micro architectural optimizations can be done using a commercial HLS tool, Xilinx Spartan/Virtex boards and Xilinx EDK design suite. The results presented in this paper show how various architectures in hardware software codesign flow can be chosen keeping energy efficiency in mind. They also show how they reduce design cycle time to reach the optimal results.

2.3.1 Summary: Low-Power High-Level Synthesis Work

Dynamic power is one of the most important components of power consumption of a design, and thus its reduction is targeted during most power-aware high-level synthesis processes. Current approaches at RTL or higher level are designer's knowledge dependent. Most of the approaches do not provide any support for power reduction from the behavioral specifications itself. In this thesis, we make an attempt in that direction. We propose approaches to enable clock-gating from the C description itself for various granularities of clock-gating such as fine grain at variable level and coarse grain at function or scope level. We also show how to extend this approach for sequential clock-gating. Finally, we present how to utilize power models to guide power reduction process at high-level. The advantage of such an approach is facilitation of power reduction features at the high-level. Also, the approaches for power estimation, which will be discussed later in detail, combined with power reduction approach can make the design flow completely at high-level. This will help in providing power aware design methodology at high-level with faster turn around time.

Chapter 3
Background

We discuss the necessary background topics, which will help in understanding this book. We briefly explain the components of average power consumption, because this book addresses the problem of power estimation and reduction at the high-level. We provide an overview of FSMD modeling using GEZEL, which helps in understanding the approach discussed in Chap. 5. We briefly provide an overview of the ESTEREL [60], PowerTheater [128] and C2R [41]. These tools are used in this book and will help the reader in understanding the details in a few chapters. We briefly discuss various modeling styles used at the high-level under Transaction Level Modeling (TLM) section. We finally touch upon the basics of clock-gating and sequential clock-gating.

3.1 Average Power Components

Average power dissipation in an hardware design consists of the following three basic components [123]:

1. *Dynamic Power* – It is related to switching activity occurring in the design. Dynamic power can be expressed as

$$\text{Dynamic Power} = kCV^2fa$$

where, k is a constant, C is the overall capacitance, V is the supply voltage of the design, f is the switching frequency of the component, a is the switching activity.

Dynamic Power is the dominant source of power dissipation in hardware design, and is highly dependent on the application and architecture. A lot of opportunities to reduce dynamic power exist at high-level. Power reduction techniques discussed in this book concentrate on reducing the dynamic power consumption of the hardware design. Switching activity and capacitance are two most important factors to keep in mind at high level because small changes in hardware architecture may impact the capacitance and activity profile of the design.

S. Ahuja et al., *Low Power Design with High-Level Power Estimation and Power-Aware Synthesis*, DOI 10.1007/978-1-4614-0872-7_3, © Springer Science+Business Media, LLC 2012

2. *Leakage Power* – It is the static component of the power dissipation and occurs due to spurious currents in the non-conducting state of the transistors. Static power has two important factors, namely *Sub-threshold leakage* and *gate to source leakage*. As process technology node size is decreasing sub-threshold leakage is increasing. This factor exponentially increases with newer technology node. Some of the prominent techniques to reduce leakage power include power-gating and stacking of transistors. These techniques are applied after the netlist is finalized for the design. Gate to source leakage can be reduced by improving the gate insulation of a node of transistor.

3. *Short-Circuit Power* – It occurs due to the stacked P and N devices in a CMOS logic gate that are in the ON state simultaneously, and can be minimized by reducing the signal transition times. It is hard to reduce this component of power through synthesis-time optimizations. After the netlist is finalized, a lot of care is given to place the design so that there is not much voltage drop for some transistors. Voltage drop on nodes may not only cause timing issues but also short circuit power issues. If short circuit power persists for longer time it may cause extremely high power consumption.

Various tools exist in industry that can perform power estimation accurately at the RTL or lower-level such as PowerTheater [128]. These tools require design information for models in Verilog/VHDL, simulation dump from the RTL or lower-level simulation (such as Value Change Dump (VCD) or FSDB) for activity analysis, and technology library. Such information helps in doing power estimation accurately and efficiently at the RTL and lower level.

3.2 PowerTheater

PowerTheater [128] is an RTL/gate-level power estimation tool, which provides good accuracy for RTL power estimation with respect to the corresponding gate-level and silicon implementation. We have used PowerTheater extensively for power estimation purposes. PowerTheater (PT) accepts design description in Verilog, VHDL or mixed verilog and vhdl. Other input required for average power analysis is value change dump in vcd or fsdb format dumped from RTL simulation. PT also requires power characterized libraries in .lib or .alf format for power analysis. Apart from doing average power estimation, PowerTheater also helps designers to perform the activity analysis for the testbenches, probabilistic power estimation, time-based power estimation, etc.

In simulation-based approach, PT first reads the design description provided in verilog/vhdl then it extracts activities from VCD and convert it into GAF (Global Activity File) and finally it maps the activities of the different signals obtained from RTL simulation of the design and does the power analysis. The most time consuming part in the simulation-based approach is activity extraction. If testbench is written for fairly large simulation time, then size of value change dump will be very big and hence activity extraction will take longer time, which will further elongate the power analysis time.

Vectorless power estimation method is used when stimulus for the design is unavailable. In vectorless approach, PowerTheater reads the design description in Verilog/vhdl, then it assumes that activities of ports and internal signal will be provided by the user in Vectorless Activity Format (VAF). Once the activity of input–output ports and internal signal (if available) is provided, PT performs the probabilistic power estimation by propagating the activity of the missing signals from the activity provided for inputs and outputs. The accuracy of this method depends upon the accuracy of the activities of input, output ports and internal signals in vectorless activity file.

3.3 FSMD Modeling Using GEZEL

The model of computation for GEZEL is hardware-oriented and is based on FSMD. GEZEL environment supports co-simulation with several instruction-set simulators as well as with SystemC and Java. In this book, we utilize the modeling style of GEZEL for FSMDs and obtain state-wise toggle count of the design. Various features and advantages of the GEZEL are available at [68], here we list a few main advantages of the GEZEL:

- GEZEL uses cycle-true semantics with dedicated modeling for control structures and allows compact representation of the micro-architecture of the domain-specific processors.
- The simulation environment is scripted for fast edit–load–simulate cycles.
- GEZEL simulation environment provides a support to collect datapath activity per cycle during the simulation. This information can be further utilized for capturing total activity per state to create statistical model for average power computation.

3.3.1 An Example

In GEZEL, there exists a separate datapath and control corresponding to each FSMD. In the control path, we describe the control behavior of the design using finite state machine. The mapping of state machine to the description is self evident in the GCD example shown in Listing 3.1. The state machine performs certain computation functions invoking datapaths based on the input conditions in each state and moves to the next state. In the GCD design, state s0 of the FSM checks if ldr signal is true and then invokes the signal flow graph "run" and moves on to state s1. The datapath "run" performs the core GCD computation.

Listing 3.1 shows the snippet of gcd example written in GEZEL. In fsm, we describe the controller behavior of the design using finite state machine. In every state, state machine performs some computation and goes to next state based on certain condition. For example, in the gcd design shown in Listing 3.1 the state s0 of the fsm checks if ldr signal is true, then perform operations in "run" signal flow graph (sfg), which does the core computation for gcd. GEZEL provides an environment where designer can express an FSMD and can do cycle-accurate simulation for the design as well. Detailed description of GEZEL is available at [68]. GEZEL supports advance datatypes such as integers etc. and helps in modeling FSMDs.

Listing 3.1 GCD code snippet in GEZEL

```
// Datapaths
sfg init { m = m_in;
n = n_in; }
sfg run { m = (m > n) ? m − n : m;
n = (n > m) ? n − m : n; }
}
// State machine capture for FSMD
fsm ctl_gcd(gcd) {
initial s0;
state s1;
// State machine information
@s0 if (ldr) then (run) −> s1;
else (init) −> s0;
@s1 if (doner) then (init) −> s0;
else (run) −> s1;
}
```

3.4 ESTEREL

ESTEREL is an imperative language for modeling synchronous reactive systems [23], especially suited for control-dominated systems. It has various constructs to express concurrency, communication and preemption, whereas data-handling follows the style of procedural languages such as C. Its semantics can be understood as a Finite State Mealy Machine (FSM), but it must ensure determinism, so that a program generates the same output sequence for the same input sequence. Internally, the ESTEREL compiler translates the control part of a program into a set of Boolean equations with Boolean registers.

3.4.1 Esterel Studio

Esterel Studio (ES) [60] is a development platform for designing reactive systems, which integrates a GUI for design capture, a verification engine for design verification and code generators to automatically generate target-specific executables. The GUI enables modeling through the graphical safe state machine specification called SSM or the ESTEREL textual specification [23]. ES performs static as well as run-time consistency checks on this model for non-determinism, dead code, etc. The next most important ingredient of ES is the formal verifier (*esVerify*), which verifies the designer's intent. It allows both assertion-based verification as well as sequential equivalence checking. User-defined properties (expressed as assertions) and automatically extracted properties (out-bound, overflow, etc.) are formally verified by generating the appropriate counter-traces (.esi files) to illustrate violations. An assertion is flagged as an *Assume* to notify *esVerify* that it is an assumption about the environment. It can be applied as a constraint during formal verification (FV) of the other non-*Assume* assertions. Finally, ES performs multi-targeted code generation, which range from targets such as RTL (in VHDL/Verilog) to ESL (in C).

3.5 Multivariate Least Squares Linear Regression Model

Linear regression is a regression analysis in which a relationship between multiple independent variables and dependent variable is modeled by least square function. This equation is called as linear regression equation, it is a linear combination of one or more model parameters, which is called as regression coefficients.

3.5.1 Theoretical Model

Consider a sample of m observations done on n variables $(X_1, X_2,...,X_n)$. If these n variables are assumed to satisfy a linear relation with the response variable Y, then it can be represented as

$$Y = \beta_0 + \beta_1 X_1 + \beta_2 X_2 + \cdots + \beta_n X_n. \tag{3.1}$$

For the regression model shown in (3.1), there can be various ways to calculate the value of regression coefficients β such that the error between the predicted and measured value of response variable is the minimum. Let's denote the ith observation on these variables as $(X_{i1}, X_{i2},...,X_{in})$, $i = 1, 2, ..., m$. Let's say Y_i is value of the response variable Y for the ith observation. *Least squares error* method can be used to reduce the difference between predicted and measured values, objective function can be represented as

$$min \left(\sum_{i=1}^{m} (Y - \beta_0 - \beta_1 X_{i1} - \cdots - \beta_n X_{in})^2 \right). \tag{3.2}$$

3.6 Transaction-Level Modeling

In this section, we briefly discuss the important terms used in the system-level domain specially related to transaction level modeling (TLM). There are a lot of references that are available on TLM, we briefly touch upon the terms and concepts explained in the paper by Cai et al. [39].

Transaction-level modeling (TLM) is used to increase the productivity of design flow. In such kind of modeling style, models are presented at the higher level and can be used for quick design space exploration purpose. In TLM, details of communication models are separated from details of computation models. Figure 3.1 presents the system modeling approaches discussed in [39]. This paper briefly discusses the models commonly used in the design flow. Figure 3.1 provides a relationship between various TLM models through the relationship between communication (y-axis) and computation (x-axis). On each axis we can see three different time approximations namely un-timed, approximate-timed and cycle-timed. Un-timed model generally represents the functional model without any timing information. Cycle-timed accuracy can be considered as fine-grain accuracy for TLMs, while approximate-timed model can be any model having timing accuracy between untimed and cycle-timed models.

Various type of models used in design flow are captured in Fig. 3.1. Specification model can be considered as the most abstract model, where computation and communication have no timing information and it is generally used for functional verification purposes. In component assembly models, computation is modeled in

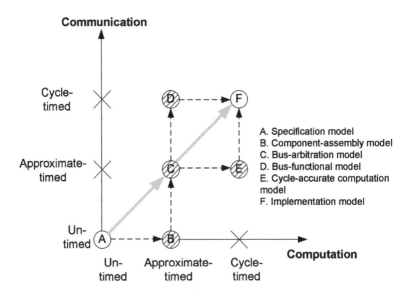

Fig. 3.1 System modeling discussed by Cai et al. [39]

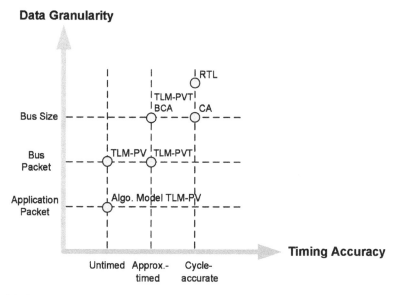

Fig. 3.2 System modeling discussed by Frank Ghenassia [70]

approximate time, while communication between processing elements (PEs) is not modeled with any timing information. Bus arbitration model represents approximate timing for the communication as well as the computation of the design. Generally the communication between PEs is modeled using channels without any cycle-accurate information. A bus arbiter is inserted into the system to arbitrate any bus conflict. Bus functional model represents the cycle-accurate communication and approximate time computation. Constraints of communication are captured using timing information or clock cycles of bus clock. Cycle accurate computation model contains cycle-accurate computation and approximate time communication. This model can be generated from the bus arbitration model, it contains more detailed computation modeling at the cycle-accurate level. Finally, implementation model generally captures PEs at RTL, communication channels are generally represented using wires. PE interface is pin accurate in the implementation model.

For SystemC various TLM models are discussed in [70]. TLM modeling shown in Fig. 3.2 is specific to SystemC, if we closely monitor modeling paradigm for SystemC. Its modeling accuracy also depends on timing accuracy and data granularity while in the earlier discussed paradigm, it was discussed more in terms of communication and computation. These paradigms discuss the need of various level of abstraction for modeling different systems and its importance at various level.

To illustrate data granularity, one can take an example of video IP. If IP is modeled using frame based algorithm, then data transfer is done frame by frame at the coarse grain level, which can be considered as application packet level.

If IP is modeled at the finer granularity then transfer can be represented as line or column based, or a transfer consisting of both line and column. While at the finest granularity it can be represented as pixel by pixel to transfer the video data. While timing accuracy varies from untimed to cycle-accurate (cycle by cycle behavior), anything falling in between is approximately timed.

The purpose of untimed TLM is primarily creating functional software model and functional verification. Since this model is created primarily for software programmers, it is also named as transaction level model with *programmer's view (PV)*. *Programmer's view plus timing (PVT)* models generally try to capture micro-architectural details containing essential timing annotations for behavioral and communication specifications. These models are generally used for checking the simulation accuracy for real time embedded software development and architecture analysis. Such timed TLMs are also known as PVT. Figure 3.2 captures the modeling accuracy of the untimed and timed TLM with respect to other modeling style such as bus cycle accurate (BCA), cycle accurate (CA), and register transfer level (RTL) models.

3.7 High Level Synthesis Using C2R

C2R [41] is a high-level synthesis tool, which takes ANSI C specifications as input and generates synthesizable RTL as output. We have used this tool extensively in our studies. Here we discuss in detail the design methodology based on C2R and salient features of this tool.

- *Specification*: As we know that for any design flow, specification is the starting point of the design flow. In the C2R flow, ANSI C specification is taken as input. Since C is one of the most widely adopted languages, it is relatively easy to obtain specifications for various important applications such as compression, image processing, security related IPs, etc. Such specifications in C helps the designer to understand the application domain better and helps in fixing design requirements at the specification stage itself.
- *Restructured Specifications*: This is one of the most important stages of the design flow, which requires the highest effort from the designer. In this stage, the designer works on the macro-architecture of the design based on the specification and design requirements and provides a restructured design. In this context, the restructuring phase includes inserting the necessary parallelism in the appropriate parts of the code. C2R provides user a way of compiling the restructured spec-ification using gcc. This helps the designer to ensure the functional correctness of the design and understanding if various ways of introducing parallelism can affect the functionality of the design.

    To enable sequencing and parallelism in the input C language, C2R requires restructuring to be performed by the designer using some directives. This restructured language is taken as an input by the C2R compiler. Restructured design from the macro-level can be understood as concurrent computation units, C2R provides the directives to include the parallelism. One can go through the code-flow (which is sequential C code) and introduce parallelism based on his needs. By default, the compiler selects a particular implementation based on control flow. To change the default flow, the designer has to provide some extra information. For example, at the highest level one can extract independent processes using the c2r_process. Inside a thread, a designer can further exploit the parallelism using c2r_fork or c2r_spawn directives, in which instead of sequentially executing the code one can exploit the inherent parallelism in the code/design and use it at process-level. C2R also provides an opportunity to go a level deeper to enable parallelism, in which the user can ask the C2R to produce the micro-architecture utilizing lesser clock or more clock based on the requirement.

Such parallelism can be exploited using forwarding of clock, this technique utilizes the local data dependency, if specified then compiler will put the clocks in the generated RTL Verilog. For all the stages, compiler has a default behavior in which it tries to find the maximum parallelism, while other variants of macro-architecture can be obtained by including directives such as c2r_fork, c2r_process, etc.

- *C2R*: There are three main stages/passes in the compiler. We will briefly touch these passes and provide an overview on how compiler understands the input provided and how it processes them.

  - *Front end Pass*: Compiler first does the lexical processing for parsing the input specifications. Front end pass can be understood as the pass in which the compiler mainly gathers the information from the restructured C and creates the abstract syntax tree (AST) for further processing. In this stage, the compiler checks the correctness of the input language and passes through the expression-typing stage to check the lexical and syntactical correctness. Also, the compiler starts building the symbol-table to record all variables, functions, declarations, etc. This symbol table used by the later passes to extract the information related to types, size, scope, attributes, etc. The compiler also performs syntax checking, type checking, etc.

  - *Middle Pass*: This pass can be considered as a pass where AST is massaged based on the macro-architecture definition provided by the designer. In this pass, the compiler does processing for loop-unrolling, linearizing (i.e. flattening constructs, such as if–then–else, while, switch, for, etc.). Every stage introduces some additional details on the AST or elimination such as dead code elimination, etc. This middle pass can also be considered as a stage where compiler goes through some optimization passes as well such as clock-insertion pass. Finally, before the backend code generation, compiler extracts the control/data flow graph (CDFG) from the AST. Finally, an appropriate finite state machine is extracted from the CDFG of the compiler for back end processing to generate the synthesizable Verilog.

— *Back end Pass*: Backend pass mainly works on the CDFG provided by the middle pass and creates the Verilog/ cycle-accurate SystemC. In this stage, the compiler finds out appropriate constructs of Verilog to be put on various nodes, which are already scheduled and allocated.

• Back end Flow: In the context of high-level synthesis, the flow which utilizes synthesizable RTL can be considered as back end flow. The generated verilog can be used in any FPGA/ASIC flow using tools provided by xilinx, synopsys, etc. There is a difference in the back end pass and back end flow. Back end pass is the pass to perform post processing on the CDFG, while back end flow is applied on the HLS generated RTL. Back end pass is internal to the high-level synthesis tool.

3.8 Power Reduction Using Clock-Gating

Dynamic power consumption remains to be the biggest contributor to the total power consumption of a hardware design. Register power consumption is one of the highest contributor to the dynamic power, making register power as one of the biggest contributor. One of the most prominent technique to reduce power consumption of a hardware register is clock-gating. In this section, we discuss the basics of *clock-gating* and its variant known as *sequential clock-gating*.

3.8.1 Clock-Gating

A typical register in hardware is shown in Fig. 3.3. For a register shown in Fig. 3.3, the input to the register is the output of a multiplexer behind it. This multiplexer output is dependent on its EN input, which is the control signal for the multiplexer. If the EN is true, then the register takes IN as input else it retains the previous value. If the EN signal does not change frequently (i.e. new data is not coming frequently to the register), the register still consumes dynamic power every clock cycle. The clock-gating gates the clock based on the EN signal, which stops the clock driving

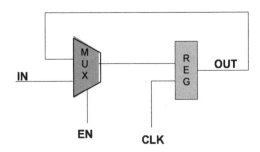

Fig. 3.3 Typical register representation in hardware

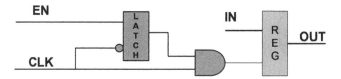

Fig. 3.4 Clock-gated register representation in hardware

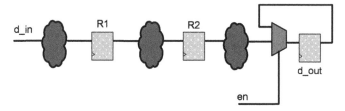

Fig. 3.5 An example of de-asserting the datapath registers to find out the sequential clock-gating opportunities [125]

the register when it is not needed. There are many ways to gate a clock, and one common way is the latch based clock-gating, pictured in Fig. 3.4. More details on clock-gating can be obtained at [58].

3.8.2   Sequential Clock-Gating

Sequential clock-gating is a technique which is applied across the clock boundaries. Sequential clock-gating is applied to gain further power savings [97, 105, 125]. At the RTL, sequential clock-gating opportunities may exist in many different ways, some of which are discussed below with examples.

3.8.2.1   De-assert Enables in the Earlier Cycles if Forward Stage Is Clock-Gated

In the example shown in Fig. 3.5, three registers d_out, R1, and R2 are connected in a pipeline. The register (d_out) has feedback path from a multiplexer. This multiplexer's controlling input is en. If en is true, then d_out accepts value from the preceding combinational logic, else d_out retains its old value. Other registers do not have any multiplexer before the input; hence, no enabling conditions to control the register value.

Suppose we find that the register R2 and its preceding combinational logic do not feed into any other part of the design except the logic preceding d_out. We also find that register R1's fanout cone does not include any other logic than shown in the figure. Now, if en is false, at cycle t, any activity on R1 in cycle t-2 (2 cycles earlier) will not affect the d_out. Similarly, any activity on R2 at cycle t-1 will not be relevant either. Discovery of these kinds of relationships between the registers

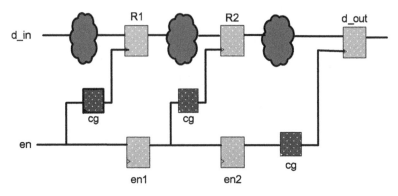

Fig. 3.6 After applying the power reduction technique on the example circuit [125]

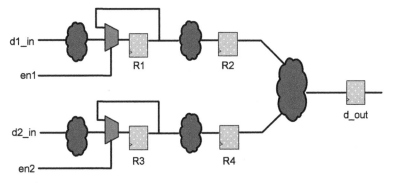

Fig. 3.7 An example of de-asserting the datapath registers in forward stage enable to find the sequential clock gating opportunities [125]

can be exploited for clock-gating registers R1, R2 to save clock-power wasted in the redundant computations. Figure 3.6 shows the power aware circuitry after utilizing the relationship for these enabling conditions. Since an enable signal for d_out already exists, RTL synthesis tool can recognize the clock-gating opportunity for d_out. On the other hand, clock-gating of R1, R2 requires the above analysis because sequential gating opportunities for these registers are not visible within a single clock boundary.

3.8.2.2 De-assert the Forward Stage of Enable in Later Cycles if the Current Stage Is Clock-Gated

Figure 3.7 presents another example where the power reduction technique discussed above can be applied. In the example circuit shown here, datapath starts from the inputs d1_in, d2_in and ends at the output d_out. The two inputs are enabled by en1 and en2, respectively, while there is no enabling condition for d_out. If there is no change in value of R1/R3 at an arbitrary clock cycle t, then there will be no change in the value of R2/R4 at clock cycle t+1, and no change in the value of d_out at cycle

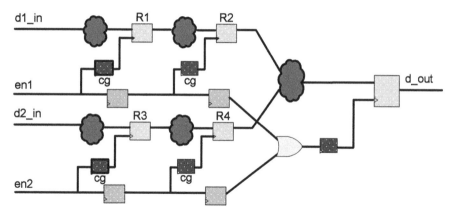

Fig. 3.8   After applying optimization in example shown in Fig. 3.7 [125]

t+2. Of course, this has to mean that the fanout cones for each of these registers are limited to what is shown in this figure. For this scenario en1, en2 can be propagated to control the clock power of registers R2, R4, and d out. Figure 3.8 shows the final design after applying sequential clock-gating optimization.

Now consider another example in the Snippet 1. If register r changes in cycle t, then register q changes in cycle $t + 1$. Moreover, if r does not change in cycle t, then q does not change in $t + 1$. This information cannot be inferred from structural information as were the cases in the previous two examples. However, a model checker can easily infer this information, and hence we can eliminate the computation of u, z (provided the only usage of u, z were creating the enabling condition for q's updation), and clock gate q a cycle later if r is gated in the current cycle. Since this is a simple enough example, one could see it as obvious. But for more complex scenarios, only a model checker or state space exploration based method can infer these kinds of information. This also gives us a motivation to further research sequential clock-gating technique and adopt it at higher abstraction level.

Snippet 1   Dynamic Opportunity of sequential clock-gating for registers r, q

```
Input int in1,in2;
wire int x, y;
register int u,z,r,q;
....
x = in1 + 1;
y = in2 - 2;
next(u) = in1;
next(z) = in2 -3
....
if (x-y > 10 ) then next(r) = expression;
....
if (u-z > 10) then next(q) = expression2;
```

Chapter 4
Architectural Selection Using High Level Synthesis

Advances in semiconductor technology still governed by Moore's law, have enhanced our ability to create extremely large and complex microelectronic systems. To utilize the increased number of transistors on a single chip, and to cope with the ever increasing demand for more functionalities in electronic systems, system-on-chip (SoC) based designs are being increasingly used. A growing trend in SoC design is to have heterogeneous processors on a single chip. In the embedded systems domain, a common performance enhancement strategy is to use dedicated co-processors on which the main processor can off-load compute intensive tasks. Compute intensive tasks such as encryption, decryption, compression, matrix mul-tiplication, etc., were earlier done using software on the main processor. Therefore, a large number of sophisticated software algorithms exist and are often available in very efficient C-based implementations. Since time-to-market windows are forever shortening, reuse has become a mantra in the industry. Why not reuse efficient, well validated, and time tested software algorithms, and synthesize automatically for power/energy, area, and latency efficient implementations? Such a path to hardware would spare the designers the excessive time required to develop the co-processors from scratch as well as reduce the resources required to verify the hardware in RTL simulation, which is quite slow. With this goal in mind, we present here C2R compiler based [1] methodology for reusing existing C-based implementation of compute intensive algorithms. We also present, how one can systematically carry out architectural exploration at the C-level and automatically synthesize efficient hardware co-processors.

The question is what are the main challenges we face in achieving this goal? First, the C/C++ implementations are targeted for sequential implementation, but their corresponding hardware has to be highly concurrent, hence requiring discovery of concurrency opportunities from the sequential implementations. Second, simple parallelization of the sequential code does not take into account resource constraints, and hence we need to consider various alternative ways of exploiting the available concurrency. Third, depending on the latency, area and power requirements, there exists various ways of exploiting the concurrency, and hence a choice has to be

S. Ahuja et al., *Low Power Design with High-Level Power Estimation and Power-Aware Synthesis*, DOI 10.1007/978-1-4614-0872-7_4,
© Springer Science+Business Media, LLC 2012

made through proper design space exploration. Fourth, once the proper architected form of the original C-code is realized , we need to verify that it has the exact same functionality as the original C-code. Finally, we need to find effective synthesis algorithms from this architect-ed C-code to efficient RTL in Verilog/VHDL.

In this chapter, we detail how we tackle each of these challenges in the C2R methodology. The process of identifying concurrency in the computation dictated by the given algorithm, and then expressing it in concurrent C-code using C2R directives is called "restructuring" in the C2R jargon. Restructuring is the process of imposing a hardware architecture on top of the computation. Of course, there are many distinct choices of architectural changes that one can make, and hence an architectural exploration cycle must ensue to find the one that meets all the latency/area/power constraints. Since the C2R compiler can synthesize hardware from the C-code, one possibility would be to try an architecture via restructuring, go through the synthesis flow, create the hardware, estimate area/latency/power and if not met, redo the restructuring. However, such a methodology would require too much time, hence the need to find ways to shorten the length of each exploration phase. In the C2R methodology, functional verification of restructured and original C-code with test benches in a high level native C environment is faster than low level event based simulation of RTL. We illustrate the entire methodology with a case study of an AES co-processor synthesis to RTL Verilog starting from a C-implementation of the Advance Encryption Standard (AES). To illustrate the efficiency of the power aware design methodology, we show the results of certain other sizable bench marks.

Main Contributions of this paper are as follows:

- Exposition of the C2R methodology and the C2R compiler for co-processor synthesis.
- Demonstration of a case study for architectural exploration for area/speed using C2R.
- Demonstration of power aware co-processor synthesis through fast architectural exploration in the C2R methodology.
- Making a case, that C-based High Level Synthesis (HLS) of co-processors may be effectively used with proper surrounding methodology.

4.1   Advance Encryption Standard (AES)

AES is a block cipher that works by encrypting groups of 128-message bits. Three different length "keys" (128, 192, or 256 bits long) can be used in the AES encryption process. The 128-bit key length was selected for use in our case study. AES operates on a 4 x 4 array of bytes (16 bytes or 128 bits) termed as state.

Fig. 4.1 Four stages of AES
encryption

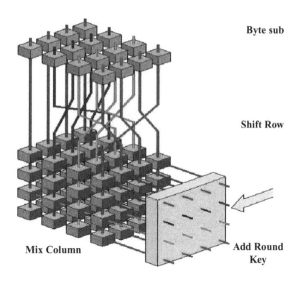

The (128-bit) algorithm consists of ten steps called rounds. For encryption, each round of AES (except the last round) involves four stages:

1. *AddRoundkey*: Each byte of the state is combined with the round key, each round key is derived from the cipher key using a key schedule.
2. *ByteSub*: A non-linear substitution step where each byte is replaced with another according to a lookup table.
3. *ShiftRows*: A transposition step where each row of the state is shifted cyclically a certain number of positions.
4. *MixColumns*: A mixing operation which operates on the columns of the state, combining the four bytes in each column using a linear transformation.

The final round replaces the MixColumns stage with another instance of AddRound Key. Figure 4.1 captures these four stages. Each round of the AES algorithm requires access to four static tables, with each table being accessed four times, resulting in a total of 16 static table reads. In addition, there are four accesses to a key table, for a total of 20 memory accesses per round. This table access characteristic of AES provides an opportunity for a number of design tradeoffs to be made between performance (throughput and latency) and area. Figure shows a simple block diagram of the AES algorithm, including the static tables and the dynamic 128-bit key table.

The C source code contains both "rolled" and "unrolled" versions of the algorithm, where the rolled version consists of a single "for loop." Each pass through the "for loop" is one round of the encryption algorithm. The rolled version was used as the starting point for what is termed the baseline restructuring of the C code.

4.2 Architectural Exploration for Efficient Hardware Generation

This section presents how various architectures can be easily visualized in the C2R methodology. We explore baseline and three optimized versions of the hardware implementation of AES Rijndael algorithm (c source code is taken from [7]) via the C source code restructuring process. The baseline represents the minimal restructuring needed to enable C2R compiler to generate a synthesizable Verilog RTL implementation of the design. We chose an application IP which is reasonable in size and is a good candidate to explain the use of C2R based design methodology for quick design space exploration. Various other examples related to quick design space exploration for design IPs such as gzip, gunzip, data encryption standard (DES) hardware blocks used for compression, decompression, and security related applications have been implemented based on C2R based design methodology [1].

4.2.1 AES Restructuring Iterations

4.2.1.1 Baseline Hardware Architecture

The baseline restructuring, based on the rolled version of the AES algorithm, introduces three interface functions (shown as queues in Fig. 4.2) to define the top-level module ports and data input/output queues of the hardware design for input data, output data and round key (rk) input. The encrypt/decrypt C function is declared with the c2r_process directive, which causes this function to be realized as a finite state machine (FSM) in RTL. The interface functions (declarations and definitions) and process declaration represent the primary changes to the C source code for the baseline restructuring. The other code changes basically consist of declarations of bit-width specific data types for C variables to minimize hardware storage requirements. As shown in Fig. 4.2, the hardware implementation contains encrypt and decrypt. Both these functions (FSMs in hardware) require interaction with static encrypt, decrypt and key table. The encrypt function contains the "for loop" that executes the ten rounds in the algorithm. Listing 4.1 shows the list of function prototypes using C2R directives.

Listing 4.1 C2R directives used for different functions

```
//function prototypes
extern uint8_t c2r_interface enc_data_in (uint8_t plaintxt, uint1_t encrypt);
extern uint8_t c2r_interface enc_data_out(void);
extern uint8_t c2r_interface send_rk (uint32_t rkkey);
extern void c2r_process encrypt(void);
```

The Rijndael algorithm requires ten large static arrays, five for encrypt and five for decrypt. By default, the static table read-only storage will compile and synthesize to wire and gates, such that the memories will be simultaneously accessible in one

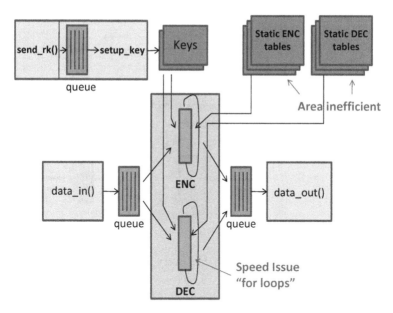

Fig. 4.2 Baseline architecture for AES

clock cycle. The round key data storage becomes flip-flops in the hardware, again allowing parallel, single clock accesses for higher performance, but at the cost of area. With the rolled version of the code, even with parallel accesses to stored data, it still requires 12 clock cycles to perform the ten rounds of encryption, resulting in the throughput of 501 Mbps as shown in Table 4.1. These numbers are obtained on Xilinx 4-inputs LUTs based Virtex-IV FPGA.

4.2.1.2 Pipelined Architecture

As can be seen in Fig. 4.2 an obvious performance deficiency exists in the baseline architecture, since a new 128-bit block of input data may not enter the "for loop" until the current block has completed all ten rounds (note that there is no pipelining in this implementation). As previously discussed, there are 20 accesses to stored data each round, so the number of clock cycles required to complete a round depends on how the storage is implemented in the hardware. This iteration of the hardware architecture is focused on increasing the throughput. The unrolled version ("for loop" unrolled) of the Rijndael algorithm was used as the starting point. The baseline suffered from the fact that a new block of data had to wait until the current block completed all ten rounds. The logical solution is to introduce pipelining into the architecture by making each round into a separate process. One such process is captured in Listing 4.2. This resulted in a ten-stage pipeline that produces a

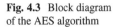

Fig. 4.3 Block diagram
of the AES algorithm

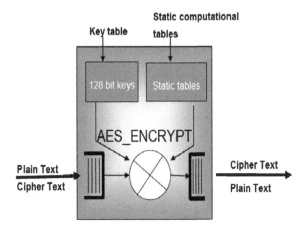

new block of encrypted (or decrypted) data every clock cycle. This new pipelined
architecture delivered a throughput of 14 Gbps, which is roughly 28 times higher
than the throughput of the baseline design as shown in Fig. 4.3.

4.2.1.3 Block Ram Based Architecture

Both the baseline and pipelined architectures allowed storage to be implemented in
gates (FPGA LUTs) and flip-flops, resulting in fairly large area, as illustrated by the
FPGA Slice/LUT counts in Table 4.1. These implementations may be suitable for
an ASIC based design, where gate and flip-flop storage would be quite efficient,
but a more area/LUT-efficient design is needed for an FPGA-based solution. In
order to achieve a lower slice/LUT count, it is desirable to utilize on-chip block
RAM resources in the FPGA. The way to accomplish this is through the use of
special accessor functions, which define the interface to hardware memories. Such
accessor functions at the C source code level instructs the compiler to generate
Verilog RTL automatically instantiating block RAMs (snippet of the code block
for the memory accessor is shown in Listing 4.3). This dramatically reduces the
number of slices/LUTs required, but the use of shared memories requires the full 20
clocks per round for data accesses, the results are captured in Table 4.1.

4.2.1.4 Performance and Area Optimized Architecture

The goal for the final iteration was to achieve optimal efficiency that is high
performance (throughput) with reasonably small area (in terms of FPGA slices/-
LUTs). This architecture used the unrolled version of the AES algorithm, pipelined
architecture, accessor functions, and a larger number of block RAMS to enable
parallel accesses to stored data for each round. Instead of taking 20 clock cycles

to access data, only 1 clock is required in this implementation. As shown in the Table 4.1, the resulting throughput is 18 Gbps and the LUT count is about one-fourth that of the pipelined architecture. (Note: Additional reductions in LUT count can be realized by further increasing the number of block RAMs.)

Listing 4.2 Source code changes applied for pipelined implementation

```
void c2r_process enc_round1 (void)
{
while (start_round1_flag == 0)
{
c2r_wait();
}
/* round 1: */
s10 = Te0[s0 >> 24] ^ Te1[(s1 >> 16) & 0xff] ^ Te2[(s2 >> 8) & 0xff] ^ Te3[s3 &
       0xff] ^ rk[4];
s11 = Te0[s1 >> 24] ^ Te1[(s2 >> 16) & 0xff] ^ Te2[(s3 >> 8) & 0xff] ^ Te3[s0
       & 0xff] ^ rk[5];
s12 = Te0[s2 >> 24] ^ Te1[(s3 >> 16) & 0xff] ^ Te2[(s0 >> 8) & 0xff] ^ Te3[s1 &
       0xff] ^ rk[6];
s13 = Te0[s3 >> 24] ^ Te1[(s0 >> 16) & 0xff] ^ Te2[(s1 >> 8) & 0xff] ^ Te3[s2
       & 0xff] ^ rk[7];
enc_run_flag = 0;
start_round2_flag = 1;
start_round1_flag = 0;
} // end enc_round1
```

Listing 4.3 Binding block RAMs of FPGAs

```
void c2r_process enc_round1 (void)
extern void c2r_foreign c2r_accessor c2r_memory c2r_useclock(clk)
     c2r_timing_model(C2R_SYNC) bram (
int c2r_width(fordepth(44)) *address,
uint1_t write_enable,
uint32_t write_data,
uint32_t read_data);
static uint32_t c2r_use(bram) rk[44]; //binds rk to bram
```

4.2.2 Results

The AES algorithm presented in this paper illustrates the ease of adoption and productivity benefits of C2R Compiler and C-based hardware design flow for ESL design. Extensive ANSI C syntax support means that existing C code bases can be compiled with minimal changes to the source, and programmers can take advantage of the full power of the C language. One can compile the input specification in ANSI C, including structures, unions, global variables, pointers to structures, complex arrays, multi-level indirection, and pointer arguments to function calls. Although one can argue that such an architectural selection can be done automatically, complete automation of concurrency opportunities is very difficult or limited to specific domain [5]. Our approach relies on the fact that architecture level decisions are controlled by the architect while the micro-architectural control is taken care by the HLS tool.

Table 4.1 AES experiments
results using C2R

AES Version	MHz	Through put Freq.	Flops put	Slices Luts	BRAMs
Baseline	47.03	501 Mbps	4,890	50,784 96,422	0
Pipelined	109.2	14 Gbps	4,662	27,760 54,068	0
Block RAM Based	126.5	80.9 Mbps	5,425	3,679 4,861	12
Speed and Area Optimized	144	18 Gbps	5,328	7,670 14,588	80

The four restructuring iterations discussed here, were completed very quickly, and thus illustrate the dramatically different results (throughput varying from 81 Mbps to 18 Gbps shown in Table 4.1). These iterations show how different architectures through small changes to the C source can be explored at the system-level. The fact that the restructured ANSI-C code can be compiled using gcc, means that verification can be accomplished in native C environment and the same source drives both architectural exploration and hardware implementation.

4.3 Functional Verification/Simulation at High-Level

4.3.1 Guidelines for Simulation

This section details some of the necessary modifications that are required to make the restructured code execute in C development environment.

4.3.1.1 Inclusion of Main ()

It is not mandatory for the C2R code to include a main () function. However, since the C run-time environment calls the main function to begin program execution, it has to be included in the source code. The main function can be used to perform house-keeping functions for the core in C environment. This function can be compiled out for C2R compiler, if it was not initially present in the restructured code.

4.3.1.2 Data Input/Output

The manner in which data is read or written will vary between the hardware core and the software execution. The interface differences will have to be reflected in the C source and compiled out for C2R compiler. The additional input/output statements can be included in the main function that was included as part of the C source.

4.3.1.3   Handling of c2r_process

One of the significant characteristic of the restructured code is the presence of parallelism due to the inclusion of processes and forks in the code. There are two possibilities by which the code can be made to execute in the software environment: threaded-C execution and a non-threaded sequential execution. Each "process" in the restructured code is mapped to a thread and the processes are executed as threads concurrently. The other alternative is to consider each process as a function and then schedule these functions in a manner that would satisfy the dependencies requirements for the processes. This would make the restructured code run in a native C sequential environment. But the designer takes the responsibility of scheduling the functions in the correct order and including the function calls at the appropriate places in the code. The calls to these functions are compiled out for C2R compiler and are only visible to the C compiler, as the processes would be executed concurrently in the hardware. Similarly, processes are coded as infinite loops in C2R. The infinite loops are compiled out for native C execution.

4.3.1.4   Possibilities of Bit Mismatch

C2R allows the designer to declare variables of all sizes while it is not feasible to map variables of all sizes in C environment. This bit mismatch could create possible differences between the two environments. Let us consider declaring a 1-bit variable in the restructured code which cannot be exactly mapped in C. Hence, the coding style when working with 1-bit variables becomes important so as to make the functional verification possible in C.

Consider the following example:

Listing 4.4 Example

```
uint1_t foo;
utin10_t blah;
.
.
.
foo = ~((uint1_t) (blah >>8));
if (foo)
{
    true_code_block;
}
else
{
    false_code_block;
}
```

The 1-bit variable foo is assigned the 2nd MSB and checked in the following if condition. This will work as expected in the hardware but foo will be an 8-bit integer in C environment and hence will be assigned the two most significant bits. But such mismatches can be easily debugged in the C environment and appropriate changes for the C environment can be included. One possible solution is provided below.

Listing 4.5 Code changes illustration

```
#ifdef C2R
if (foo)
#else
if (foo == 255)
#endif
```

4.3.1.5 Pointer Initialization

In a C2R environment, it is possible to bind a pointer to a custom memory accessor and then access the memory elements with the index. By doing so, the pointer need not be initialized in the restructured code.

Consider the example:

Listing 4.6 Example showing binding to pointers

```
static some_struct  c2r_bind(some_custom_accessor) *some_ptr;

// And memory elements are accessed by using

some_ptr[index];
```

However, the C environment expects the pointer to be initialized and this has to be considered by the designer. Moreover, if the memory accessor gathers the data from different memory banks and assembles the data, it would be difficult to initialize the pointer to one array. In such cases, the statement some_ptr[index] can be mapped to a function in the C environment using a macro. The same memory accessor can be used as the function that gathers the data from different arrays.

Listing 4.7 Changes for the example shown earlier

```
#ifdef C2R
static void c2r_memory c2r_accessor some_custom_accessor (address, write_enable
    , write_data, * read_data )
#else
   static return_type_of_read_data some_function (address)
#endif
{
Collect data from different memory elements; //body of the memory accessor
}
```

4.3.1.6 Comparison of Structures

C2R allows direct comparison of two structures without the need to compare the individual members of the structures. However, a C compiler would consider this as an error and the source needs to be modified for the C environment. The use of memcmp can essentially replace the comparison of structures as in the following code segment.

Listing 4.8 Structure example

```
some_struct x, y;
#ifdef C2R
  if (x ==y)
#else
  if (memcmp(x, y, sizeof (x))
#endif
```

When the possible trouble spots are considered and taken care of by the designer, the restructured code can be compiled and executed in a functionally equivalent manner to the hardware. The advantages of having a functionally equivalent C source are detailed in the following section.

4.3.2 Advantages of C-based Verification Model

The significant advantage of having a functional software model is the reduction in execution time. C-based execution of the functional model will be several orders of magnitude faster than corresponding RTL simulation. This advantage can be exploited by testing the model against an innumerable number of test vectors as possible. This faster execution also helps in cores where the model can be reconfigured in a number of ways depending upon the design parameters. An extensive permutation of design configuration parameters might not be possible using hardware simulations. But all such possible combinations of various design parameters can be tested in a short order of time using the functional model. The data collected using such runs can be used to characterize the relationship between the design parameters.

Besides the faster execution time, the C-based model can also exploit the tools and utilities that are available in the C development environment.

4.3.2.1 Code Coverage

The same restructured code is used to generate the hardware and also for executing the functional model. Hence, a code coverage performed in the C environment will have a correlation to the statements executed in the hardware platform. The gcov tool can be used in conjunction with gcc to perform code coverage. This tool can help in source code analysis and discover where optimization efforts are needed. Untested segments of the code can also be discovered using this tool and test vectors can be generated to test the corner-case code segments. Thus, this tool also helps in the development and improvement of testsuites, which can later be used during hardware simulation. In addition, this tool also provides information as to how often each line of code is executed and the number and probabilities of each branch being taken.

4.3.2.2 Functional Coverage

Functions that perform functional coverage on selected modules can be added to the source code and calls to these functions can be compiled in only during debug process. The coverage information can be useful in determining whether all sections of the selected modules are being accessed. The gcov tool along with the functional coverage code can provide this information. If certain sections are not being used, the coverage information can then be used to modify the size of the module that was being tested, if needed. Conversely, the coverage information can also help to characterize input test vectors. An example coverage code segment is provided below.

Listing 4.9 Example for functional coverage

```
#define HASH_SIZE 4096
#define HASH_BIN_SIZE HASH_SIZE/8
void hash_coverage (int index)
{
    if (index > 0 && index <= (HASH_BIN_SIZE-1))
    {
        bin0++;
    }

  if (index > HASH_BIN_SIZE && index <= (2*HASH_BIN_SIZE-1))
    {
        bin1++;
    }
  if (index > 2*HASH_BIN_SIZE && index <= (3*HASH_BIN_SIZE-1))
    {
        bin2++;
    }

    {
        other coverage bins;
    }
}
```

4.4 Conclusions

In this chapter, we show how starting with an efficient software implementation of a compute intensive task, one can perform architectural exploration in the native-C environment, and synthesize hardware co-processors quickly. We briefly provide an overview of C based verification models and power-aware micro-architecture exploration using clock-gating. It shows that staying in the native C-environment for architectural exploration combined with compiler based power saving technique insertions can lead to fast design of co-processors. These co-processors have been shown to speed up many computations by orders of magnitude [12]. For a true ESL development methodology, it is very important to work at different levels such as the untimed C level, focusing on the objectives ranging from verification to hardware development. Our example and approach in this paper shows how one can efficiently perform architectural/ micro-architectural exploration. We believe that even if HLS cannot be used for all general purpose hardware synthesis, certain specific kinds of hardware synthesis can be achieved with proper HLS and surrounding methodologies.

Chapter 5
Statistical Regression Based Power Models

Accurate power estimation is one of the most important ingredients for any successful design methodology in embedded system or system on chip (SoC) design. Ability to estimate power accurately and early in the design flow helps in controlling form factor, battery life and size, etc. Current power estimation techniques are well established at the RTL and lower levels of abstraction, while the techniques above RTL suffer from accuracy problem. One of the main reasons for such inaccuracy of power estimation at higher levels is the lack of technology specific information and implementation details of a design. Power estimation at the RTL or above, requires design information in the form of Verilog/VHDL/System-C/C model, technology libraries, simulation dump, etc. However, the more detailed the model is, the slower the simulation and hence the slower the estimation. In order to make this process faster it is desirable to use power estimation techniques that do not require all this information directly exposed in the model. In the past, various techniques have been proposed to achieve this goal. These techniques involve capturing the activity by simulating a high-level model and feeding it to the lower level power estimation techniques [9, 11] using the library of blocks containing power related information [34], and using efficient power models/macro-models to estimate the power consumption [113, 119], etc.

The focus of this chapter is on creating an efficient power model to estimate the power consumption of a particular design/design-block without burdening the simulation model with all the design details and technology related information. The goal is to maintain a balance between the estimation speed and the accuracy achieved. In this work, we target a cycle-accurate FSMD based modeling style and figure out what information must be collected from their simulation to obtain sufficiently accurate power estimation without having to simulate the RTL of the design in a full-chip simulation. Therefore, we experimented and verified the model's accuracy on 90-as well as 180-nm technology libraries to show the applicability of our model. The cycle-accurate FSMD level of abstraction is less detailed than RTL. This FSMD model can be co-simulated with other FSMD models, and even C-models.

S. Ahuja et al., *Low Power Design with High-Level Power Estimation and Power-Aware Synthesis*, DOI 10.1007/978-1-4614-0872-7_5,
© Springer Science+Business Media, LLC 2012

This provides the benefit of power aware exploration at the highest possible level of abstraction avoiding an RT level full-chip simulation. Even though the power consumption of an implementation depends upon the technology libraries and other implementation details; signal toggle plays a statistically significant role in power consumption. So, if the signal toggles at the various states of the design can be statistically correlated to the power consumption over a sizable number of simulation examples, one can learn the dependence of various statistics available at the FSMD level on the actual power consumption of the implementation. Since learning is aided by a detailed model of the design with specific technology library information, such information is folded into the regression model in the methodology presented in this chapter. We will refer this methodology as Statistical regression based power models for Co-processors Power Estimation (SCoPE) hereafter in this chapter.

SCoPE methodology utilizes the activity/toggle information of each state and corresponding datapath explicitly available in the FSMD modeling style. Note that in the RTL model, the distinction between state and datapath activities for specific states is not explicit. As a result, state specific toggle counting is harder. The FSMD abstraction comes to the rescue here. Learning based on the activity of the design at its various states is done by inferring a regression equation relating such statistics of the abstract FSMD model and the power consumption of the implementation model. At the end of such a learning phase, the power model is trained/learnt, and it can be used for estimating power consumption based on state specific activity information extracted from the co-simulation of the FSMD and the rest of the chip (other parts of the chip could also have FSMD substitutions for various components). In the *SCoPE* methodology power numbers can be obtained just after co-simulating the co-processor models with the simulation model of ARM processor.

There are quite a few advantages of *SCoPE* methodology: (a) it utilizes the explicit partitioning of design in states and datapath provided by the FSMD modeling style (which helps in extracting relevant run-time information), (b) once the model is ready it requires only activity/toggle information per state (which can be obtained from the simulation of the FSMD) to get an estimate on power consumption, and (c) these models can directly be attached with the simulation model of the processors. Main features of *SCoPE* methodology can be summarized as follows:

- A two-phase power estimation technique for components in an SoC design using multi-variate regression utilizing a cycle-accurate FSMD model of the design.
- A compositional methodology for power estimation of SoCs utilizing the statistical power models of its components while simulating at a higher abstraction level.
- Experimental demonstration of the accuracy of this technique for some co-processor examples.
- Demonstration of a flow based on this methodology which incorporates GEZEL [68] for FSMD modeling and activity capture, PowerTheater [128] for learning, and JMP [81] for regression modeling.

5.1 Regression Based Power Model for FSMDs

The core idea is based on the fact that activities associated with a particular state of the FSMD model corresponds to the specific part of the implementation model as shown in Fig. 5.1. Note that methodology discussed in this chapter is applicable to the language/design-flow capable of representing cycle-accurate FSMD models such as Esterel [60], SystemC [116], etc.

Let's consider an FSMD **M** with state set **S** for a test run for T clock cycles, α_s is the total number of times design visits state **s** during the test. Let τ_i^s be number of signal toggles during the ith time the design visits state **s**, then total number of signal toggles in state **s** can be represented as

$$\Gamma_s = \sum_{i=1}^{\alpha_s} \tau_i^s \tag{5.1}$$

While the design is in state **s** during the test, energy spent due to toggles in state **s**, $E_s = k\Gamma_s C_s V^2$, where k is a proportionality constant, C_s is the capacitance, and V is the operating voltage. The total dynamic energy spent in design is assumed to be the sum of energy spent in each state and can be represented as

$$E_{dyn} = \sum_{s \varepsilon S} k\Gamma_s C_s V^2$$

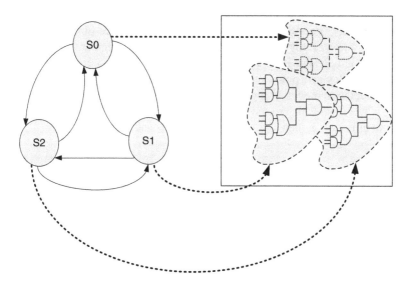

Fig. 5.1 Conceptual mapping of activity to the real design

Now, if clock frequency of design is f then the average dynamic power can be represented as:

$$P_{dyn} = E_{dyn}/(T/f)$$
$$= \sum_{s \varepsilon S} k\Gamma_s C_s V^2 f/T$$
$$= \sum_{s \varepsilon S} k(\Gamma_s/T)C_s V^2 f$$
$$= \sum_{s \varepsilon S} K_s(\Gamma_s/T), \text{ where } K_s = kC_s V^2 f$$

Total power of a design is sum of dynamic and leakage power, then the equation for average power can be represented as:

$$P_{total} = P_{leakage} + \sum_{s \varepsilon S} K_s(\Gamma_s/T) \tag{5.2}$$

In (5.2), K_s is a constant for state s under the assumption that the state-wise toggles at FSMD level are proportional to the toggles occurring on the implementation model. K_s can be obtained by taking the ratio of the dynamic power numbers obtained from RTL estimation to toggle count information obtained from FSMD simulation and number of clock cycles for which the test was run. Our Experiments with 90-*** and 180-nm technology libraries show that leakage dependence is captured in this model and power numbers obtained using this model are close to RTL power estimation numbers.

Relationship between toggles and total power, maps well to the regression model discussed in Sect. 3.5. Variables of the regression model are obtained by taking the ratio of state-wise toggles and simulation duration in number of clock cycles. Regression coefficients can be obtained by applying least squares error objective on these variables and power number obtained using RTL power estimation on the implementation model. Using this model, we can reduce a lot of complexity in creating power models/macro-models because most of the time these models require a lot more information than toggle/activity of inputs and outputs, etc.

5.2   Steps for Power Modeling

There are two models involved in this power modeling methodology: the FSMD model and the power model. The FSMD model is a cycle accurate abstraction of the implementation model. This model is the input for our power modeling methodology. The power model captures a relationship between the toggle statistics of an FSMD model and power estimation done on implementation model. This regression model is the output of our power modeling methodology.

Fig. 5.2 *SCoPE* methodology for creating and utilizing power models of FSMDs

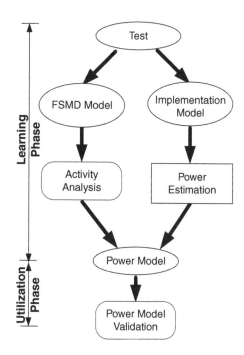

5.2.1 Learning Phase

Following steps are needed to perform learning phase of the *SCoPE* methodology (also shown in Fig. 5.2):

- *Test setup*: *SCoPE* methodology utilizes statistical modeling tool JMP [81]. Based on the experience of expert users, it requires at least 30 experiments for creating accurate statistical model using JMP. In this study, we found that regression coefficients for power model converged to a stable value for 80 tests. Random tests are utilized in our experiments. As we know that power consumption is very much dependent on the input vectors/simulation, the better stimulus provided for the analysis the better power model (more accurate) we obtain. There exist quite a few approaches in the research where techniques to find the appropriate input vectors based on the hamming distance, data correlation, etc., are presented [99]. However, the main focus of *SCoPE* methodology is on creation and utilization of the power model. The stimulus generation approach presented in [99] can also be utilized in our framework for training power model during test setup stage of the methodology.
- *FSMD Model*: We use FSMD models written in GEZEL. An FSMD is a cycle-accurate model of a controller with a datapath [155]. FSMD modeling in GEZEL clearly distinguishes between control and datapath operations. Modeling style for datapath and control is very different but that is captured well in GEZEL using state transitions for controller and expressions on signals or registers for

datapath operations. FSMD model should capture the information on every state and corresponding datapath.

- *Activity Analysis*: In this step, we mainly concentrate on obtaining the toggle count associated with every state. GEZEL simulator provides a facility to get a toggle count for every datapath unit for each cycle. For getting activity count associated with every state, we instrumented the GEZEL code and processed its simulation output using a c-shell script; this collects the sum of toggles for every state for the whole simulation duration. This activity information is used for training and validation of the power-model. GEZEL environment is capable of doing activity estimation associated with each datapath unit. In the script, we made sure that for every cycle each datapath unit can be associated to its corresponding state. This information is further processed to get the count of total toggle for every state during the simulation.

- *Implementation Model*: We use an RTL model written in Verilog as an implementation model for our analysis. One can also use the gate-level or even more detailed model in *SCoPE* methodology. We performed the analysis using RTL power estimation numbers. We assume that the finite state machine for the RTL model is same as the one used in FSMD model. RTL contains more detailed information on the datapath and implementation of the hardware design. Hence RTL model is assumed as a reference point for our analysis. Power numbers are measured from the RTL power estimator PowerTheater [128], keeping in view that such an estimator provides fairly quick and accurate results for RTL circuit as compare to its silicon implementation.

- *Power Estimation for the Implementation Model*: Power estimation for the implementation model requires design information (Verilog description), activity information (we use Value Change Dump (VCD) generated by the RTL simulator) and technology library (we use 90 and 180 nm power characterized libraries). Power estimator provides the average power numbers corresponding to VCD of every training vector. As we go lower in level, the power estimation will be very much time consuming but accuracy of power numbers obtained will increase and this will further help in improving the accuracy of regression based power model.

- *Power Model*: As discussed in Sect. 5.1, we use regressions for creating the power model, JMP [81] is used in our framework for creating regression model. We utilized activity information from activity analysis step along with the measured power information from the power estimation step for training the model. We used model fitting feature of JMP using multiple regression model and objective as least squares error to get accurate value of the regression coefficients of the power model (as discussed in Sect. 5.1).

5.2.2 Utilization Phase

Validation of regression model is very important and we perform the validation of power-model during utilization phase.

- *Power Model Validation*: Once the regression model is ready, we validate the power model using random tests. In this stage, we check how closely the output produced by the predicted model matches to the power numbers obtained from the power estimation stage discussed earlier. Most of the reported work use root mean square (rms) error (such as [34,113]) which can only be helpful for average power over various tests. In our analysis, we are calculating error for each test which gives us insight on worst case scenario (rms error is less than the worst case error).
- *Usage*: Total toggles for every state is given as an input to the power model. Power model provides a quick estimate on power consumption using regression equation and does not require power estimation at the implementation level.

5.3 Results and Conclusions

5.3.1 Tool Flow

We have used PowerTheater [128] for power estimation, VCS [149] for RTL simulation, GEZEL [68] for FSMD modeling, and JMP [81] for creating regression model. Here we discuss in detail about the tools and wrapper we built for the *SCoPE* methodology. Figure 5.3 shows the tool flow used in the *SCoPE* methodology. Grey divider line shows different phases. GEZEL simulation environment, c-shell script is required in both phases. From the GEZEL simulation environment, toggles for every datapath units can be obtained. The C script wrapper collects toggles per state for the whole simulation duration and associate the appropriate toggles to each datapath computation unit. In the learning phase, we provide the Verilog RTL model, value change dump and technology library information to PowerTheater for power estimation. These power numbers along with the state-wise toggle information are then passed to JMP for regression modeling. Once the model is ready the power estimation is performed in the utilization phase using the regression model (as discussed in Sects. 5.1 and 5.2).

5.3.2 Experimental Results

We applied *SCoPE* methodology to various designs shown in Tables 5.1 and 5.2. In our study, designs have several states varying from as low as four different control states in the EUCLID design to ten different control states (one for each of the ten rounds of encryption) in the AES design. Complexity of the designs in terms of their size, ranges from 1,099 NAND gates (for EUCLID) to 47,166 equivalent NAND gates (for AES). Tables 5.1 and 5.2 capture power numbers, worst case, and RMS error for various designs. Figures 5.4 and 5.5 provide a comparison of power

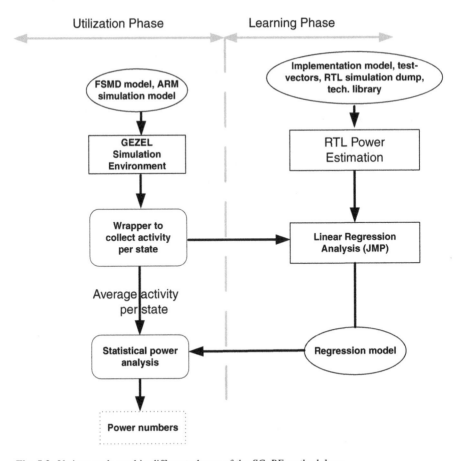

Fig. 5.3 Various tools used in different phases of the *SCoPE* methodology

Table 5.1 Comparison of measured power (90 nm) and predicted power for various designs

Design	Pred. power (mW)	RTL power (mW)	Worst error (%)	RMS error (%)
AES	145	138.99	4.14	2.26
EUCLID	0.497	0.500	0.66	0.387
UART transmit	0.485	0.525	8.28	4.15
XTEA cipher	0.530	0.538	1.58	0.607
XTEA decipher	0.541	0.551	1.99	0.78

numbers for different simulations of our approach with the PowerTheater for 90 nm and 180 nm respectively. Experimental results show that our model shows a good accuracy for different technology nodes.

Table 5.2 Comparison of measured power (180 nm) and predicted power for various designs

Design	Pred. power (mW)	RTL power (mW)	Worst error (%)	RMS error (%)
AES	455.3	468	2.79	1.49
EUCLID	1.86	1.98	6.09	3.81
UART transmit	1.59	1.68	5.67	2.95
XTEA cipher	1.765	1.75	0.86	0.25
XTEA decipher	1.79	1.827	2.08	0.79

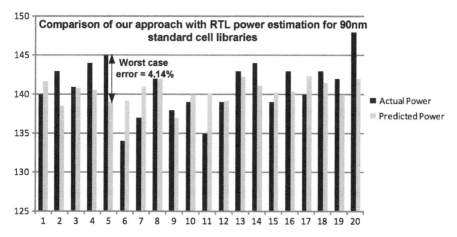

Fig. 5.4 Comparison of measured power and predicted power for AES at 90 nm

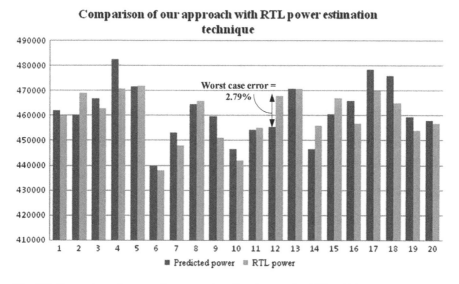

Fig. 5.5 Comparison of measured power and predicted power for AES at 180 nm

5.3.3 Discussion

5.3.3.1 Comparison with State-of-the-Art RTL Power Estimation Techniques

Power estimation utilizing power models presented in this chapter provides the following advantages over the simulation based RTL power estimation methodology:

- Once the power model is ready, it can be utilized for accurate power estimation requiring FSMD simulation only.
- Power model presented in this chapter is a simple regression model which does not require going through the overheads involved in RTL power estimation such as processing RTL simulation-dump, technology libraries and design information, etc.

5.3.3.2 Factors Affecting Accuracy

Accuracy of such power models is dependent on various factors. These factors include test vectors, training of regression models, etc. Test vectors are very important because if the model is created using the test vectors that are being used for most common usage scenario then the accuracy of the power model can be further improved. In our analysis, we have used random vectors to train and validate the power model. Earlier related works [113] report RMS error which do not provide per sample average power accuracy and is not a good measure for average power analysis per sample/test vector. As we know that RMS error is less than the worst case error, so for the cases where one needs RMS error our framework will provide even better results. Training of regression model can also be further improved by doing the training on more test vectors. Finally, the regression model we have used is multi-variate least square error based regression model, which is more suitable in reducing the average deviation of power consumption. Such kind of models may not be suitable for the designs with very high power variation for different states.

Secondly, in our power model leakage dependence is also captured with the same regression model. We utilized the power numbers from the implementation model to train the power model (in our case RTL model of the hardware design). The power numbers obtained from the implementation model include the numbers coming from state dependent leakage as well. In the proposed power model, state dependent leakage numbers are not taken into account, which has further affected the accuracy.

5.3.3.3 Tools, Environment and Speed Up

We utilize GEZEL, JMP and PowerTheater tools for experimentation. We anticipate such a methodology can be used with other framework/tools capable of representing

FSMD. Simulation environment may require some additional work to obtain the toggle count associated with each state and corresponding datapath of the FSMD. As it is visible in our approach, power estimates can be obtained at the end of GEZEL simulation. In this approach, we have attempted to increase the abstraction level to cycle accurate FSMD and still maintained a sufficient accuracy. We can remove a part of design cycle (that is the need of lower level power estimation), which leads to reduction of possible delays in the design cycle. To give an idea, we collected the time spent for RTL simulation and power estimation of AES design; which is found to be around 11 min of CPU time (7 min for power estimation and 4 min for RTL simulation for a test vector of almost 40,000 clock cycles). While in our case it is only around 5 min of CPU time. We believe that if such an equation based model can be inserted in the C source code of system-level model, it can show very good speedup. We have inserted a power model in the SystemC model in the different context [14] and found that the power models can provide a very good speedup (upto 2–3 orders of magnitude). However, such an analysis is not performed here.

5.3.3.4   Scalability of Our Framework/Methodology

In this chapter, we presented our approach on co-processors used for acceleration of certain data intensive task. In our experiments, we found that design with ten or more control states at FSMD level takes almost same or lesser execution time as compared to RTL. Learning phase for AES design with ten states and 80 samples (test vectors) took less than 5 min of CPU execution time. Most of the time in learning phase was spent in GEZEL simulation, once the state specific toggles were collected then regression model was created within a few seconds. Hence, we think our approach can be scaled to the co-processor of reasonable size. In our case, we utilized existing framework to extract state specific datapath activity information from the GEZEL simulation, while such feature might not be available with other frameworks. Other frameworks based on C/SystemC will require some instrumentation of design source code or environment.

5.3.3.5   Impact of RTL or Lower Level Power Reduction Techniques

Power model presented in *SCoPE* is trained using RTL power estimation method. RTL power estimation tools may not provide the impact of power reduction techniques such as clock-gating, power-gating, etc. Now a days power estimation tools such as PowerTheater [128] can perform optimizations such as clock-gating, operand isolation at RTL abstraction of the design. In such a scenario, it is recommended to train the power model after applying power reduction. To verify the claim we trained the power model for xtea cipher using power estimations performed after applying clock-gating on the design. We found that for the same test vectors worst case error was 1.55% and rms error was found to be 0.56%.

Similarly, multiple voltage domains inside a co-processor are not considered in the *SCoPE* methodology. However, this methodology would still be useful in case a user wants to use multiple voltage domains at the granularity of co-processors. This type of power model would require more learning if a lot of power reduction is obtained by techniques which are available after gate-level design stage. Such optimizations include leakage power reduction techniques using stacking to reduce the power consumption.

In this chapter, we presented a means to solution using GEZEL, JMP and PowerTheater. We assume such a methodology can also be used with other framework/tools capable of representing FSMD. Such a methodology is capable of saving various iterations, which design/EDA engineers might have to go through otherwise to obtain the power numbers in their design flow.

Chapter 6
Coprocessor Design Space Exploration Using High Level Synthesis

6.1 Introduction

The traditional computing arena of hardware/software co-design has gained importance in the context of system-on-chip design methodologies. Apart from the classic issues of partitioning, communication and granularity, the need for quick estimation of metrices such as area, power, and latency has turned out to be important. This can be attributed to the large design space under consideration. The number of IPs integrated on a system is large and each IP may have multiple configurations. Many of these IPs are often hardware coprocessors that accelerate compute-intensive tasks. A collection of coprocessors chosen based on Pareto optimal points w.r.t. speed, area and power may not necessarily add up to corresponding system level Pareto optimal points [67]. One way to address this issue could be through developing algorithms or heuristics that lead towards Pareto optimal configuration. Obtaining a generic strategy, keeping in mind global optimization, is hard. Furthermore, evolving mathematical models for such diverse systems is very difficult. Another approach involves using partial or full simulation of systems for different configurations. Various standard co-design platforms are available for the same [69, 146]. However, exhaustive design space exploration will be difficult.

Suppose it is required to configure a system involving (1) a data filter that processes the input data, (2) a computing core that performs a series of data-intensive transformations, and (3) an encryptor that encrypts the data before transmission. As already stated, picking the fastest IPs will not necessarily lead to the best solution. One has a plethora of choices when it comes to the filter. Multiple criteria will need to be considered to maximize throughput at the least cost. The computing engine for data transformations has to be as fast as possible. However, one might have multiple choices for interfacing this coprocessor with rest of the system. Also, the communication bottlenecks might mask the speed gains obtained from this coprocessor. Such a process will not only cause wastage of resources (in terms of power, area, etc.), but also unnecessarily increase the design time. When it comes to the encryptor, apart from multiple encryption algorithms to choose from,

S. Ahuja et al., *Low Power Design with High-Level Power Estimation and Power-Aware Synthesis*, DOI 10.1007/978-1-4614-0872-7_6,
© Springer Science+Business Media, LLC 2012

one can perform microarchitectural exploration, for example, between a pipelined version and a simple implementation.

In this chapter, we show how HLS can be used for addressing some of the problems discussed so far. HLS involves synthesizing hardware from high level language descriptions. Firstly, when using off-the-shelf IPs, instead of doing RTL design exploration, one can use behavioral descriptions and perform quick analysis in HLS. Secondly, hardware prototypes can be created quickly for software development. This aids concurrent hardware and software development. Finally, if one goes for C-based synthesis itself, modular designing will be easier at behavioral level than at the RT-level, which involves a lot of microarchitectural detailing. Individual blocks within a design can be subjected to different levels of refinement without affecting other blocks.

We demonstrate a methodology that enables quick design space exploration through prototyping for FPGA-based platforms. We use C2R to synthesize hardware description from C-based designs. Such a methodology provides a fast and easy way to explore the design space. We present a series of case studies of varying complexity in which we start from a high level description in C language, and then follow a series of C2R specific restructuring, implement the synthesized RTL on the FPGA platform and obtain the performance numbers. A variety of implementations and architectures are tried out to arrive at different trade off points. In doing so, bulk of the changes are performed within the C environment and thus design cycle time will be reduced. We also perform power estimation for the generated designs.

The main contributions of this work include:

1. Exploring possibility of using HLS for co-design
2. Platform dependent optimization at high level
3. Analyzing system level performance parameters in coprocessor selection

6.2   Related Work

We now present a brief overview of how these issues have been tackled traditionally. There are 3 distinct approaches one can come across. The first method involved developing coprocessors at RTL for each possible configuration and then selecting the best suited choice. One could go for synthesizing on platforms along with corresponding software or may resort to behavioral simulation of coprocessors and the software separately. Many industrial tools are available to support such RTL simulations including Modelsim [103], VCS [149], etc. Similarly, for the software, the simulators are available for many general purpose processors. But, evaluation of hardware/software interfaces cannot be done in such methods. Furthermore, the design cycle is too large and making small changes in the configuration will involve high design effort. This method is more suited for prototyping in ASIC flow before the design freeze and this cannot be scaled to SoC designs.

Yet another recent development involves use of behavioral languages for this purpose. A wide range of languages and corresponding cosimulation environments have been developed considering various aspects of hardware/software co-design. This includes Bluespec [29], Handel-C [44], Gezel [69], etc. Bluespec comes with an exhaustive toolset encompassing a compiler, simulator and its own IP library. The toolset takes in specification in the form of a specific language called Bluespec system verilog and generates synthesizable RTL. In another closely related work, Sullivan et al. [146] perform similar activity using Handel-C for algorithm description. This language is based on Occam. They use DK design suite from Celoxica for the benchmarking. This methodology claims to provide considerable advantage for creating communication ports, establish interfaces, etc., with ease. Finally, Gezel is a cycle based hardware description language based on finite state machine and data path. It provides a cosimulation environment and also generates synthesizable VHDL. With this, one can perform power-performance analysis and make various trade offs [72]. All of these methods reduce the verification time and also reduce many of the issues concerning hardware/software interfaces. However, they suffer from a major drawback – the need for using a new language and/or the design environment. This would mean longer design time and investment.

There is a class of ESL tools that try to leverage the software development methodologies for hardware synthesis – CatapultC [101], C-to-silicon [37], and C2R [41]. These tools take high level language inputs like C/C++/SystemC, etc and generate synthesizable RTL. Of course, the standard inputs will need to undergo a process of restructuring to make them more hardware compatible. However, one can extend or limit the level of abstraction and corresponding detailing – the restructuring activity can be limited to just providing tool specific keywords and converting data types to appropriate forms to being able to provide cycle accurate information at behavioral level. As we show in this paper, many explorations can be made and trade offs drawn through HLS. Although many of the HLS tools, including the one we have used for our experiments [4], have demonstrated how powerful hardware can be obtained and almost comparable to hand codes, our aim here is to highlight the prototyping advantages.

6.3 Methodology

The traditional co-design methodology for FPGA is shown in Fig. 6.1. It starts from the system specification which encompasses all the functional requirements and the design constraints. The behavioral C model is the most favored choice. This is followed by a profiling of the system and then doing the hardware/software partitioning to meet all the design constraints at minimum cost. The hardware specification is then implemented through the RTL design flow using HDLs and the software specification is realized in a C/C++ environment. The hardware design undergoes various levels of refinement to meet the essential area and power constraints, and to settle all the timing issues while the software flow involves functional

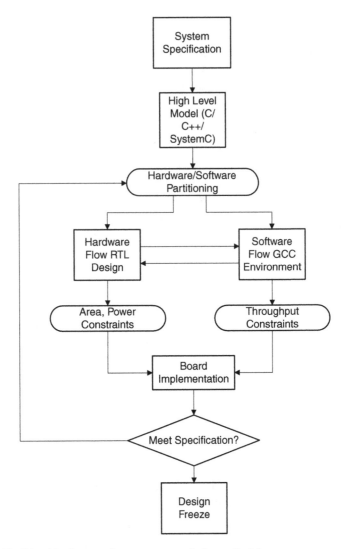

Fig. 6.1 Traditional hardware–software system co-design methodology

verification and throughput considerations. Various cosimulation methodologies are used for the functional verification and simulations. And then, the implementation is transformed to FPGA and other hardware targets. The system obtained is checked for all system level constraints and specifications. Such a cycle will have to be repeated multiple times before the final design freeze. Amongst the many drawbacks of such a methodology, the cosimulation takes the prominent role. Compared to hardware emulation-based approach, cosimulation is inherently slow, many details are abstracted and hence is not very accurate.

Fig. 6.2 HLS-based high level design flow

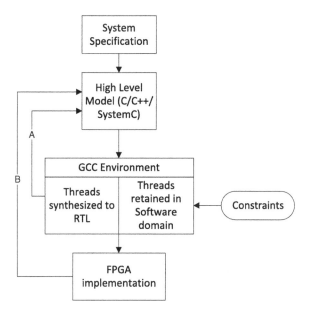

In contrary to the approach discussed above, the HLS-based approach we propose is shown in Fig. 6.2, where integrated development of hardware and software is possible. Our framework takes threaded C code as input and provides a plethora of opportunities to make dynamic hardware–software partitioning, moving components between hardware and software quickly and finally allowing complete functional verification at each stage. A set of compute-intensive threads in the C specification are identified and synthesized into hardware. The system performance parameters are analyzed and bottlenecks are identified. If the timing constraints are not met, more threads will have to move to hardware. If the area constraints are not met, some of the components may have to be moved back into the software. If both the constraints are not met, multiple threads may have to be moved between hardware and software. Since the synthesis of hardware happens very quickly and the tool automatically generates hardware interfaces, many combinations can be tried out within a short time. Also, mapping algorithms can be applied to shortlist the possible combinations.

The inner loop A in Fig. 6.2, will be frequent and quicker. The outer loop B will be longer as in the traditional case; however, it will be infrequent. The loop A involves obtaining area and speed estimates. The area estimates determine the cost of moving a specific thread to hardware. One can decide what specific components need to be moved to hardware without affecting performance significantly. The speed estimates help identify key communication bottlenecks. However, certain estimates like power analysis would still need outer loop involving FPGA implementation.

6.4   Case Studies and Results

We demonstrate the implementations on Xilinx Spartan-3e board [161]. The platform meets all necessary requirements for hardware–software design space exploration through its microblaze softcore for sequential executions and the reconfigurable logic for hardware implementations. Microblaze is a RISC-based softcore provided by Xilinx [162]. C2R tool was used as high level synthesis tool for all the case studies. We dynamically determined the threads that need concurrent execution and turned them to hardware. The remaining threads were run on microblaze.

The area, in terms of slice count and the speed up are the key parameters taken into consideration. We estimated the power for IPs in the context of system design using Xpower tool available with Xilinx ISE environment.

6.4.1   Fibonacci and Caesar

We start off with an illustrative Fibonacci IP that gives out the Nth Fibonacci number that is input to the hardware. The high level specification in restructured C version consists of a driver that gives the data input to the Fibonacci thread. This thread computes the Nth Fibonacci number and returns the output back to driver thread. We retain the driver thread in C to be run on microblaze and realize the Fibonacci function on hardware as an IP core. This IP core is attached to the softcore through PLB bus. There are four ports for the IP – the call input that activates the hardware, the data input port for the hardware, the data output port for the result, and finally the acknowledge port to indicate completion of the operation and thus closing the handshake. These four ports form a boundary between hardware and software and we create a memory mapped register corresponding to each of them. In the software environment, we define these address locations as volatile pointers that can be directly accessed for read and write. This provides good scope for performing behavioral verification of IP in C environment.

Another IP that we worked on was Caesar cipher that involves streaming data flow framework. In this IP, the data to be encrypted is continuously sent to the computation unit. Depending on control signal, the computation unit either encrypts or decrypts the data and sends back the data to the processor. There were three threads in the design: the cipher thread which did the actual computation, the data input thread for giving data and data output thread for collecting the output data. While the computation thread was realized in hardware, the I/O threads were retained on processor.

The results are tabulated in Table 6.1. The speed up in column 2 of Table 6.1 is the ratio of number of execution cycles for coprocessor-based implementation to the execution cycles on software implementation. The area in column 3 is given in terms of the slice count. The speed up number obtained for the designs is

Table 6.1 Performance matrices for fibonacci and caesar

IP	Spd	Area	Sys Pwr SW(mW)	Sys Pwr HW(mW)	IP core Pwr(mW)
Fibonacci	13	2,652	686.7	685.2	0.13
Caesar	27	2,675	701.3	697.4	1.51

quite small. Although the raw processing power of hardware is very high, Amdahl's rule takes over most of the gain in terms of overall design. The speedup obtained in Caesar cipher is slightly higher. This is because we are using a stream of data here rather than a single byte as in the previous case. The speed gains obtained from hardware implementation can be justified only when the amount of data processing exceeds the communication overhead of hardware/software co-design. The HLS-based design flow can help in making fine grained decisions in such scenarios efficiently as the design time is less. Various combinations of hardware/software components can be tried for a given data pattern. With regard to the area, microblaze itself occupies about half the total number of available slices, which means that the area of coprocessor block is about a tenth of the total system area tabulated here. Nevertheless, overall slice count had to be included as we are considering system parameters.

As in the case of performance figures, the IP core power alone cannot be the point of focus. Column 4 in Table 6.1 refers to the total system power in software while column 5 refers to system power for coprocessor-based implementation. Column 6 gives the power of individual IP core. It's only a fraction of overall system power. We had expected the software power consumption to be lesser than that for the hardware implementation. On the contrary, the overall system power remained more or less same. So, although the IP core is turned off, the savings one gets is only limited. In smaller designs, the power numbers for software are in fact slightly higher. But for a large design like AES, it behaves on the expected lines. Though the power numbers for the software and hardware versions are similar, the hardware implementations turn out to be more efficient. The hardware will be more energy efficient owing to the speed gains of the design. We will have about 13 times energy saving for fibonacci and 27 times for caesar cipher. Thus, depending on the power and performance values, one can make fine grained decisions.

6.4.2 Bitcounter IP

In this case study, we consider an IP that counts number of 1s in a stream of 16K array. Such an IP finds use in areas like parity checking and cryptography. The data is sent 32 bits at a time to a function which counts the number of 1s. The result obtained is accumulated for all the 16K array of data. We came up with a threaded version of the C reference implementation. We formed two threads – a driver thread that initializes a 16K sequence of random numbers and calls the countbits thread

Table 6.2 Performing
platform dependent analysis
for BitCounter IP

IP	SPEEDUP	AREA	Power(mW)
Basic version	243	2,870	680.29
FIFO	276	3,212	684.57
DMA	13,402	3,072	717.01
FSL	152	2,653	682.39

which does the computation. At this stage, one can verify the functional correctness of the threaded version in C environment. After this, we did restructuring of C code. The countbits thread will be hardware implementation while the driver will be in software. The first implementation as shown in Table 6.2 is a basic implementation like the previous ones and uses PLB bus and memory mapped registers. Later, we added a FIFO to the Bus–IP interface. This was expected to increase the speed as hardware need not be tied to the slower processor. However, the speed improvement was small as only data output from IP will be enhanced by FIFO. Data coming into the IP will still be delayed by the processor. So, we used DMA to eliminate data path from the processor. Now, the hardware can get data at higher speed while the processor can do other jobs. Finally, the bus architecture itself was removed and point-to-point Fast Simplex Link (FSL) was used to assess the speed-area obtained.

The countbits implementation had a higher data size of 16Kbytes. The software version was slow and also the communication overhead had been greatly overcome because of the data size. This explains the big speedup in column 2 of Table 6.2. A FIFO-based implementation, as expected, increased the speed by a certain extent. But this was at the cost of extra slice count as seen in the area numbers. The FSL implementation, gives a relatively smaller speed up. One can consider using such an option for hard-deadline IPs that need dedicated communication channels. FSLs are highly optimized paths for Xilinx environment and hence occupy small area.

The power consumption for overall system for coprocessor implementation is given in column 4 of Table 6.2. The software baseline implementation consumed 694 mW. While the FSL and FIFOs marginally affect the total system power, DMA has a much bigger influence. FSL kind of a dedicated hardware resource gives significant performance gains at a very small power cost and hence can prove effective in many situations.

Presented in Table 6.3 is a comparison between implementation of IP core using verilog[1] and C2R for DMA implementation. Although the hand code version is faster than C2R code, the overall system speed is same for both the cases. They are limited by the speed at which DMA can transmit the data. This makes a strong case for high level approach. There could be design scenarios like this, wherein we don't need high speedup. In this case, it's good enough if the IP can finish its computation before DMA gives new data. Such decisions are difficult to be made at RTL. Despite a slightly higher area of the C2R generated core, we see that the power numbers of the hand coded IP and C2R generated IP are relatively same.

---

[1] The verilog code was implemented by Mike Henry, ECE, Virginia Tech.

Table 6.3 Hand code versus
High level code for DMA
implementation

IP	SPEEDUP	AREA	Power(mW)
Hand code	13,402	74	0.70186
C2R code	13,402	91	0.71701

Table 6.4 Different
implementations of AES IP

IP	Start time(nS)
Basic version	27,410
Pipelined version	36,070

6.4.3 AES

We consider the implementation of advanced encryption standard (AES). AES is
a block cipher-based encryption standard [127]. The core consists of mainly four
functions: adding roundkey, substitute bytes, shift rows and mixcolumns. There
are 11 rounds in all. In its simplest form, one could make a single thread that
implements all three functions of all the 11 rounds into a single big thread. Although
very inefficient, such a design can be accomplished in minutes. This can help in
performing quick estimations. However, the area of the hardware obtained turned
out to be too high for board implementation. One of the major bottlenecks for us was
the large memory that was required to store round keys in hardware. Going back to
the restructuring phase, we made a few modifications in the C code. This involved
adding directives that explicitly indicated the compiler to port the RoundKey data
into BRAMs. The next stage for this core was to introduce pipelining. The 10
rounds were divided into individual stages by unrolling the loop in the C code and
introducing pipelining signals.

Presented in Table 6.4 are simulation results from ModelSim for standalone IP.
By start time, we mean the time first AES encrypted data starts appearing on the
output data bus. One could see a slight delay in start time of pipelined version, due
to addition of handshake signals in each state and corresponding delays introduced.
The BRAM version of the core was generated and because of the area issues, only
this was implemented on the board. The core gave a modest speed up of about three
times. While the overall system power for software version was 1271.24 mW, the
coprocessor version had 1,345 mW. The AES core contributed about 122 mW to the
system power.

The need for looking at overall system performance metrics instead of individual
IP blocks has been demonstrated in [72]. They show how an individual core that
gives a speed up of orders of magnitude would only increase the performance five
times when integrated in a full system. We highlighted similar findings in the high
level approach. Incidentally, one of the cores used in demonstration of [72] happens
to be AES core, although the two are different implementations and one-to-one
comparison cannot be made.

Thus, we were able to obtain relative performance metrics similar to different
methodologies, remaining at high level for many of the time consuming activities.

Also, many high level synthesis tools have been proven to generate powerful hardwares that are comparable to hand generated cores. But this comes at a cost of increased design time. In fact, [4] demonstrates how one can make suitable changes in C code to obtain substantial speed and area savings using C2R on the AES core.

Chapter 7
Regression-Based Dynamic Power Estimation for FPGAs

7.1 Introduction

Design space exploration for SoC is a two-stage problem. First, it involves finding quick and accurate estimation methodologies to obtain the design space parameters. Next comes developing algorithms to exhaustively and efficiently search the multi-objective space. The increase in number and complexity of the IPs integrated per system has exerted tremendous pressure on both the estimation as well as the algorithmic aspects. There is a need for good estimation techniques in the context of heterogeneous system design. In particular, doing detailed and accurate power estimation continues to be time consuming. Recently, a few interesting techniques have been developed for doing early estimations with reasonable accuracy. Statistical learning techniques are one of them. But they are limited both in scope and versatility.

Dynamic power estimation has been studied well in the context of individual IPs in ASIC flow. Relatively, FPGA dynamic power estimation has not received much attention [51, 79, 130]. We focus in this chapter on linear regression-based power estimation technique for FPGAs. Current techniques develop power models for individual IPs with input and output toggles as the parameters for regression. The regression coefficients are arrived at, with extensive training of the model on sample test patterns. The power model is developed for each IP separately in terms of a linear equation relating dynamic power to the I/O toggles. But each IP will have a separate power model. New model will have to be developed even for small changes in the design. In this work, we evaluate the possibility of a unified power model for all the IPs in a system instead of developing power models for each IP separately. The random logic in ASIC IPs make the effect of I/O toggle very pronounced. On the contrary, FPGAs are structured. All designs will be mapped to a LUT-based fabric or a fixed set of FPGA resources like multipliers and Block-RAMS. We also show in our results that the effect of I/O toggles for a particular design is small and that there is a significant dependence of dynamic power on the resource utilization. We take into account all these aspects in our analysis to develop FPGA power models. To the

S. Ahuja et al., *Low Power Design with High-Level Power Estimation and Power-Aware Synthesis*, DOI 10.1007/978-1-4614-0872-7_7,
© Springer Science+Business Media, LLC 2012

best of our knowledge, this is first work which addresses generic power models for IPs, taking I/O pattern variation and resource utilization into consideration.

Currently, RTL tools like XPower from Xilinx are used. XPower is used for post implementation place and route designs and it takes simulation data in either SAIF or VCD formats. Similar tools are available from other vendors. Such tools have a good degree of accuracy as they will be fully aware of the internal structure of the resources as well as the design details like toggle rate etc from simulation information. However, the overall flow is very time consuming. VCD file generation forms the key bottleneck. Obtaining simulation data for a set of test vectors can easily take up multiple days even for small-to-mid sized designs. Also, one needs to have synthesized RTL that is ready for bitstream generation in such flows. This may be too late for big designs. These form the primary motivation for developing high level power estimation methodologies for FPGAs.

There are certain interesting commonalities in IP usage across the industry for SoC design. IP reuse and customization is very common. The generic IPs, either developed in-house or obtained off-the-shelf, would be generally subjected to extensive profiling and hence a large bank of test vectors and the simulation data in different formats will be available. The changes for each system may involve modification in design spec, IP interface differences or variation in the input data pattern. The change in design spec could alter a few configuration registers, etc., or may address larger issues like memory bank size, etc., or bring a new IP architecture itself. The changes in IP interfaces could be due to the bus protocols or architectural changes in interfacing IPs. The data patterns may vary for each class of applications and corresponding benchmarks. For all such changes in every IP, big or small, one needs to go through the stage of estimation multiple times, followed by exploration algorithms phase to arrive at an optimal system configuration.

Finally, the profiling in general, is carried out for individual IPs in isolation. The profiling information can change drastically when the IP is interfaced in a processor-based system [72]. Hence, such methods do not help much in design space exploration. The speed and power specifically, can vary significantly when measured at IP level and at system level [94]. In fact, there can be noticeable difference in system level speed and power values even for small changes in the design and I/O pattern. We, therefore, adapt system level profiling in our estimation models.

The main contributions of this work are

1. Regression modeling for FPGAs considering I/O toggle and resource utilization.
2. Unified power model for multiple IPs.
3. Comprehensive system level power estimation approach for IPs.

7.2   Related Work

Regression-based RTL power models have been studied well over the past decade. One of the early works involved developing power macro models for primitive components at RTL [77]. This work discusses in detail about the power macro-models and the overheads involved, sampling and regression-based power models. Reference [33] includes an exhaustive discussion on the different kinds of RTL power models. In this work, the power estimation in done in multiple phases, involving an offline characterization-based on I/O toggle statistics and online tuning to improve accuracy. These works focus on generic ASIC power models at RT-level taking gate-level simulation as the golden model while this methodology works at a higher level of abstraction taking RTL simulation as golden model. There were also some works which used these techniques for FPGAs [79]. Here, the authors propose an adaptive regression method to model FPGA power consumption. However, all of these works develop power models for each RTL components or IPs separately while we work towards a generic power model involving multiple IPs.

Reference [130] and [51] use an orthogonal approach and characterize power-based on the utilization of various resources of the Xilinx board. Reference [130] estimates effective capacitance of each of the major resources through transistor level modeling and estimates the switching activity using a large set of benchmarks. They combine all the data to evaluate percentage breakdown of dynamic power consumption of different components of Virtex-II FPGA chip. Reference [51] performs similar activity for Spartan-3 Xilinx board. Firstly, such a characterization for individual chips considering design information as secondary would not be very useful for cost metrics evaluation. Also, compared to these approaches, our methodology has the advantage of giving system level power estimate for IP which will be more meaningful for design space exploration. Instead [80] developed regression modeling at the operand level (like adders, multipliers, etc). Apart from the problems discussed about earlier methods, this kind of an approach can have accuracy problems at a complete design level, as one needs to make a detailed analysis on the effect of combining these operands in different ways, on power consumption.

Reference [7] presents a "Grey-box approach" to high level power estimation in ASIC flows. They consider the toggle counts in individual states of a design at FSMD level, instead of counting I/O toggles. This would aid in better accuracy. Reference [52] proposed a single power model for different FPGA power components using non-linear regression and to that extent is similar to our work. However, it has been found in our experiments that having single equation for any design on a particular board, will result in much higher errors. More importantly, the model is static and doesn't take into account the I/O toggles. This means the model gives the same power estimate for an IP irrespective of the data pattern. One can simply use tools such as Xilinx power estimator [159] to obtain approximate power quickly.

7.3 FPGA Power Models Using Linear Regression

The independent variables considered for power models would generally be boundary information like I/O toggle. A single hardware circuit is considered and the actual power consumption is measured for a sequence of inputs. The power

measured and the toggle count on each port are used to evaluate the coefficients of
the equation. Thus an equation is developed to estimate the power consumption of a
circuit. Such strategies have been developed for both ASIC and FPGA platforms in
the past. However, our experiments revealed that, for smaller circuits, the dynamic
power consumption of a circuit varied by small amounts with the toggle count. This
is expected behavior for FPGAs which have LUT-based architecture. The logic is
mapped to Lookup tables as against the standard cell mapping in ASICs. So, this
gives rise to the possibility of finding a generic regression functions for multiple
circuits. This means, once the equation is obtained for a set of training data, it
may be used to estimate the power consumption of a new circuit. Of course, new
variables will be needed to account for varying circuit sizes. In this work, we use
FPGA resource utilization as additional variables.

The dynamic power consumption for a circuit is given as

$$P_{dyn} = \frac{1}{2}V^2FC_{eff}S \tag{7.1}$$

where V is the circuit voltage, F the operating frequency, C_{eff} is the effective
capacitance, and S represents input switching activity. There are many variations
to this equation. In some cases, the switching activity and capacitance are combined
as "switching capacitance," while in some FPGA models, the utilization factor of
resources is considered as a separate entity in itself. The voltage and frequency
for all circuits will be constant in a given board. Of the two remaining variables,
switching activity is expected to be accounted for by I/O toggles, while the
capacitance by the resource utilization. So, we first develop an equation for power
consumption varying just the toggles and keeping the other variables fixed (7.2).
Then the area is varied and a new power equation is arrived at (7.3).

$$P_1 = K_0 + K_1S \tag{7.2}$$

$$P_2 = L_0 + L_1C_{eff} \tag{7.3}$$

However, the relationship of switching activity and capacitance is multiplicative
for power and linear regression cannot adequately model it. So we take logarithm of
the two power estimates and perform linear regression on them:

$$\log(P_{dyn}) = I_o + I_1\log(P_1) + I_2\log(P_2) \tag{7.4}$$

7.4 Methodology

In this section, we discuss the overall methodology used for creating the power
estimation models. As explained in Fig. 7.1, each IP is run with multiple data
patterns. The corresponding VCD files and I/O toggle reports are generated for
each pattern. The VCD file is used for estimating dynamic power from XPower.
The process is repeated for many IPs and additionally, the resource utilization

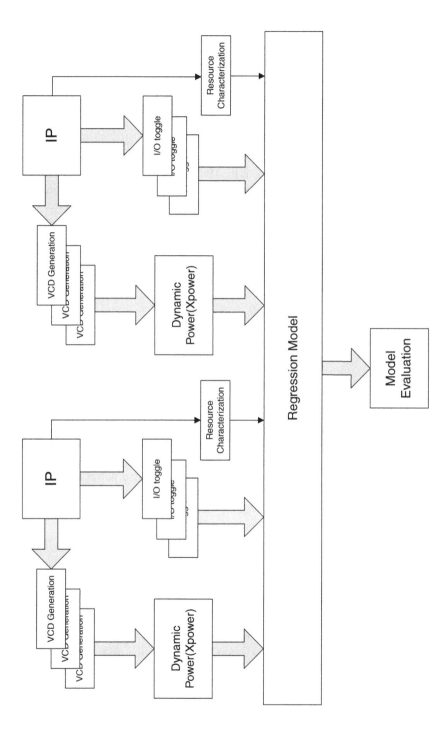

Fig. 7.1 Regression methodology for dynamic power estimation

information of each IP is also obtained. The dynamic power from XPower, toggle count for each data pattern and the resource utilization of each IP, together form the input to the regression tool. We discuss in detail each of these steps, starting with the nature of IPs selected to VCD generation phase and the statistical modeling.

7.4.1 Nature of the IPs Used

We considered 13 IPs in all, for various experiments performed. All the IPs were at synthesizable RT- level. While most of the IPs considered were "hand-coded" in Verilog, a few IPs were synthesized from a High level synthesis tool through the behavioral description of IP in C language. In order to get a broader picture, we consider the system power as our criteria rather than IP power alone. Xilinx Virtex - 2 Pro was the design board used for all our experiments. We interfaced each IP with the microblaze softcore in Virtex board through the standard PLB bus interface and then made the measurements and simulations. This gives more accurate timing and power information in the context of SoC designs.

7.4.2 Dynamic Power Evaluation for Reference Model

As is the practice in regression-based prediction modeling, we divided each sample set for different experiments into two parts. The first part was used to train the power model and the second part to evaluate the model itself. For the first part, standard power measurement flow was used. This involved performing timing-based simulation of the design using Modelsim and the simulation data was saved in VCD format. This formed the input to Xilinx XPower tool [160], along with the design files and physical constraints file. The XPower gives a detailed power breakdown based on the resource utilization and simulation information for all Xilinx boards. The static power estimate given by XPower, for a given board is constant. Hence, we eliminate that and consider only dynamic power in all our analysis.

7.4.3 Toggle and Resource Utilization Information

Apart from the actual power numbers we get from XPower for the training set, we also need information on toggles and resources used. The toggle count involved counting all the bit flips on a particular port for the duration design is run [77]. There was no particular distinction made between input and output ports for the regression model. The number of ports varied from 2 to 12 for different designs. In the regression models where different designs were considered, the ones with lesser number of ports were padded with 0. The resources considered for the analysis are multipliers, ram blocks, the FPGA slices, clock multipliers, and I/O buffers. This information was obtained from the Xilinx synthesis tools. While performing evaluation, I/O toggle information across ports can be obtained from behavioral models itself. Also, there are techniques to estimate the resource consumption from high level models [52].

7.4.4 Statistical Analysis Using JMP

We used JMP [81] for creating linear regression models in our analysis. The I/O toggle information and resource utilization information along with power information from XPower tool formed inputs to the tool. Model fitting feature of JMP was used and least squares error was the objective considered.

7.4.5 Model Evaluation Phase

This step involved finding the accuracy of our models. Various permutations of training and evaluation set were considered to arrive at better estimation models. The I/O toggles and resource utilization information for the models in evaluation phase (not given as input for curve-fitting) were plugged into the power model generated by JMP. The average and maximum difference between the actual and the estimated values formed the main criteria for model evaluation.

7.5 Results

Multiple experiments were performed to demonstrate the validity of our model, quantitatively demonstrate the improvements in our model over existing approaches and to explain how accuracy can be enhanced. The results are classified into following three subsections as under:

1. Obtaining a power model for a given set of IPs and using them to estimate power for varying datapatterns
2. Getting a generic power model to estimate power for new IPs as well as varying datapatterns
3. Grouping strategy to improve accuracy of the estimation model

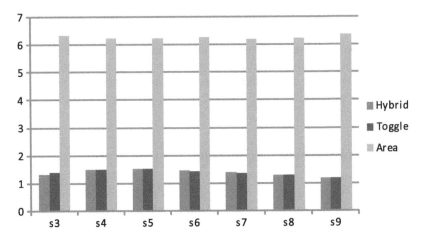

Fig. 7.2 Comparison of average percentage errors with varying data samples

7.5.1 Varying Data Patterns for Fixed IP Set

In this set of experiments, a set of seven IPs and ten different data patterns for each IP were considered. For each experiment, the number of data samples in the training set was gradually increased. Apart from estimating power from the proposed approach, which we call as "Hybrid approach," we also estimated power by taking only I/O toggle into consideration and separately with only taking resource utilization aspects into consideration. Note that the resource utilization-based model would give same power estimate for an IP, irrespective of the data pattern variation. Regression-based prediction modeling would generally require around 30 samples to form a useful model. We start with the case of 21 samples (7 IP x 3 data patterns), called s3 to indicate three data patterns considered for each IP. We go up to the case of 63 samples in s9.

Presented in Fig. 7.2 is the average percentage errors measured for each sample. Note that, the data set included in evaluation for error calculations are different from the one in training set. Also, shown in Fig. 7.3 is the maximum percentage error measured in each sample size. It can be seen that the toggle-only model and our hybrid model both give almost similar kind of errors. This is expected behavior since we do not introduce any new IPs in the evaluation set. Had the new IPs been introduced, the variation that would occur in the resource utilization will not be accounted for by the toggle-only model. The resource utilization-based model suffers as it gives same power estimate for an IP with different data patterns. This clearly shows the need to take I/O toggles and corresponding VCD-based power estimation into consideration, to get higher quality results. Even though average errors remain unchanged with the increase in sample size, the maximum error from the model drops down and stabilizes around s7. We consider s7 as the sample

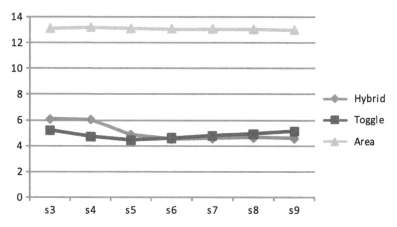

Fig. 7.3 Comparison of maximum percentage errors with varying data samples

size for all other experiments. One of the designs considered – DCT, gave very abnormal results and did not fit well into most of our experiments – we deal with this separately.

This kind of an analysis would come to good use in real-world scenarios where the SoC components are fixed and one needs to make power estimates for different application scenarios. Instead of generating VCD files for new set of benchmarks, one can plug in the I/O port information and get the power estimates. If the regression model has been developed with sufficient data patterns (like s7 in our case), one can also establish with reasonable confidence, the maximum possible deviation from the actual values. This depends on how far statistically the new data patterns are from the ones in training set. There are many techniques in literature that show how to increase accuracy and the confidence level in such scenarios [79].

7.5.2   Varying IPs for Fixed Data Pattern

In this set of experiments, we increased the number of IPs in both training and evaluation set. The number of data patterns given in each case was held fixed at seven per IP. This experiment helps evaluate the proposed model more comprehensively, as this is a generic model for IP power estimation. In the first sample space 7 different IPs were considered. Of these, six were put in training set in each experiment (and hence the name IP6) and one in evaluation set. For each IP in the evaluation set ten different data patterns were considered and the average error was calculated over all of them. Later, the sample size was increased to IP9 and IP12 and similar experiments were performed.

Figure 7.4 shows average errors measured over different data patterns for 6 IPs in the sample spaces IP6, IP9 and IP12. As, the number of IPs increase, one can notice

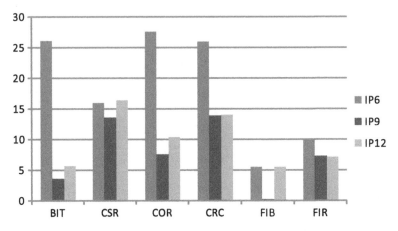

Fig. 7.4 Comparison of average percentage errors with varying IP samples

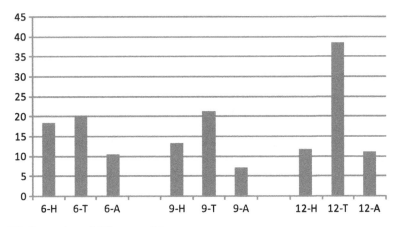

Fig. 7.5 Comparison of different models

that the percentage error goes down. However, some of the percentage errors, even in IP12 case are above 10%. This is because, the number of IPs in training set is less and the model is not trained enough for resource utilization related parameters. It is expected that the trend shown in these experiments will continue as the number of IPs increase. In real-world scenarios as well, it would be unrealistic to expect such a large number of IPs for a system. We look at ways to address this problem as well.

Shown in Fig. 7.5 is again a comparison of our model with toggle-only and resource utilization-based approaches for all three sample spaces. The vertical bars represent average errors for all IPs considered in a particular sample space and not just the six designs shown in Fig. 7.4. As expected, toggle-only approach would show maximum deviation while resource utilization-based approach performs very well. In fact, resource utilization-based approach performs better. This is because, the number of samples with varying toggles was high and the one with resource

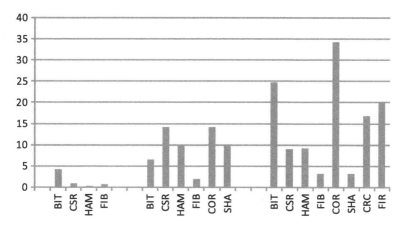

Fig. 7.6 Error increase with diverse IP sets

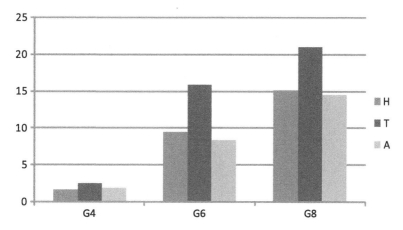

Fig. 7.7 Comparison of different models

utilization information was less. As such, lesser weights were assigned to resource utilization related parameters in equation. This trend again shows that increasing number of IPs will aid in increasing accuracy.

7.5.3   IP Grouping to Increase Accuracy

The overall accuracy with regression will not be satisfactory unless the sample space is large enough. We also noticed that in each experiment, there were certain designs which did not fit well in any models (like DCT). Analysis of the input data sequences and average errors revealed a clear pattern. The designs which did not fit were statistically very different from training sequences (toggle pattern, number of ports, resource utilization, etc). One possible strategy which addresses both these issues is grouping. In these set of experiments, we grouped IPs with similar characteristics together and evaluated error varying both data pattern and IP samples.

In the first group of IP samples, four IPs were considered. These IPs had same number of control and data ports and also were similar in resource utilization. For each experiment, an IP not included in training set and the data patterns of remaining IPs of the group, not in training set were considered. Error percentage fell sharply for all IPs (Fig. 7.6). The circuit named "Bit" shows relatively more error because of the data pattern. One of the ports was 32-bit size in this case, while all other designs had only 16-bit ports. The group named G4 had most homogeneous composition and hence least average error. The one labeled G8 had the very diverse set of IPs and hence accuracy deteriorates. Figure 7.7 shows a comparison of the hybrid model with toggle-only and resource utilization-based models.

Chapter 8
High Level Simulation Directed RTL Power Estimation

8.1 Introduction

In this chapter, we present a *high-level power estimation* methodology, which is based on a high-level synthesis framework and supports sufficiently accurate power estimation of hardware designs at the high-level. For early and accurate power estimation, the proposed methodology utilizes RTL probabilistic power estimation technique controlled by the high-level simulation. Furthermore, our methodology does not require a designer to move to the traditional RTL power estimation methodology, thus facilitating easy and early power analysis and aiding the cause of adoption of high-level design practices in ASIC design flow. This chapter provides detailed description of our methodology including tools used, algorithm for extracting activity from high-level value change dump and finally mapping this information for RTL power estimation.

Current high-level power estimation methodologies are based on:

1. Relative power estimation [158]: *Relative power estimation* at high-level helps the designer to perform early relative trade-offs for power and performance but does not provide accurate power numbers, rendering such a methodology to be unsuitable for handling design problems where accuracy is a concern.
2. Power-model reuse [20, 93]: In a *power-model reuse* based methodology, power numbers are borrowed from power-models either available from the previous generation of the chip or created from scratch. Since creating new power-models involves building a large infrastructure, such a power estimation methodology is suitable only for design problems where power-models of high accuracies are already available.

Thus, an ideal high-level power estimation methodology involves building a complete infrastructure with facilities to perform accurate power analysis and optimization. Such a methodology will target design problems that need accurate power numbers but do not have the scope of utilizing the already existing power-models. However, such a methodology is difficult to build because: (1) high-level

S. Ahuja et al., *Low Power Design with High-Level Power Estimation and Power-Aware Synthesis*, DOI 10.1007/978-1-4614-0872-7_8,
© Springer Science+Business Media, LLC 2012

design flows and tools have not sufficiently matured and (2) infrastructure required to enable the high-level power estimation is extensive – that is where our technique helps.

In the industry, accurate power estimation is currently performed mostly at the RTL and gate-level. Typically, RTL power estimation techniques require following inputs: (1) design description in Verilog/VHDL or other hardware description language, (2) RTL simulation trace in the format of Value Change Dump (VCD), Fast Signal Database (FSDB), etc., and (3) power characterized libraries (e.g. standard cell power libraries) to correctly estimate the power consumption of a design. However, it is not feasible to utilize these techniques for high-level architectural exploration and optimization because such techniques and flows are targeted to be specifically used by the RTL design teams. On the other hand, a good high-level power estimation methodology should allow a high-level designer, architect or modeler to quickly and accurately gauge the effect of various high-level modifications on the power consumption of the design without being required to completely move to RTL power estimation flow. Such a methodology will significantly reduce the complexity of the power estimation process at the high-level.

8.1.1   Our Approach

In this chapter, we propose a high-level power estimation methodology which reuses an RTL power estimation technique guided by the high-level inputs for providing reasonable power estimates. In this sense, the proposed technique assimilates the best of both the worlds and provides a feasible and efficient solution to the high-level power estimation problem.

In this approach, inputs are similar to an RTL power analysis methodology with the exception that the RTL simulation trace of the design is replaced with its high-level simulation trace for accurate power estimation at ESL. One requirement to enable such a flow is that a mapping must be established between the variables in the high-level model of the design and corresponding signals (such as those at the port boundaries) in its RTL implementation. For such a mapping to exist, we recommend utilizing a high-level synthesis (HLS) engine.

8.2   Rationale for Our Approach

In general, total power analysis time T_{total} for a design is given as,

$$T_{\text{total}} = t_{\text{d}} + t_{\text{a}} + t_{\text{p}}. \tag{8.1}$$

In 8.1:

1. t_d represents the time taken by the power-analysis tool in extracting the design information. This includes the time spent by the tool in elaborating the design and generating its intermediate representation.
2. t_a represents the activity extraction time taken by the tool for collecting power analysis related information for various signals of the design from a simulation dump (like a *VCD* file). This time is proportional to design size and simulation duration. Algorithm *VCDExtraction* corresponds to this part of the power analysis process.
3. t_p represents the time taken by the tool in performing various power calculations based on the power characterized libraries, activity values and other design information (such as fanouts).

A design can be considered as a top-level module, which instantiates m child modules. We assume that each child module consists of m child modules and h denotes the hierarchical depth. Assume that the design is simulated for t_{sim} duration. Now, consider the following notations:

n_s – Number of variables at high-level per module.
n_i – Number of intermediate signals.
t_{s_a} – Activity extraction time at high-level.
t_{s_p} – Power calculation time at high-level.
T_{s_total} – Total power analysis time at high-level (based on high-level simulation).

Corresponding notations at RTL can be denoted as n_r, t_{r_a}, t_{r_p} and T_{r_total}. Moreover, transforming a model from high-level to RTL involves the introduction of additional intermediate signals which are required to represent the model at RTL. Such intermediate signals are represented as n_i.

Thus, at the high-level t_{s_a} will be proportional to m, h, n_s and t_{sim}, and can be given as

$$t_{s\_a} = k_s * n_s * m^h * t_{sim}, \text{ where } k_s \text{ is a constant.} \tag{8.2}$$

Similarly, for the RTL, the corresponding equation can be given as,

$$t_{r\_a} = k_r * n_r * m^h * t_{sim}, \text{ where } k_r \text{ is a constant.} \tag{8.3}$$

Let's assume that on an average every variable at high-level can be represented in β bits. Thus, we can establish the following relationship between the number of high-level variables and RTL signals,

$$n_r = n_s * \beta + n_i. \tag{8.4}$$

Time taken in extracting the design information t_d can be assumed to be similar at high-level and RTL. Thus, using (8.1) at the high-level, we get,

$$T_{s\_total} = t_d + (k_s * n_s * m^h * t_{sim}) + t_{s\_p}. \tag{8.5}$$

$$T_{r\_total} = t_d + (k_r * n_r * m^h * t_{sim}) + t_{r\_p}. \tag{8.6}$$

In (8.1), power calculations time t_p is proportional to the number of signals of the design. However, for long simulation durations, activity extraction time is usually the bottleneck and is much larger than t_p. This is because size of the simulation dump increases drastically with the simulation time, and extracting information from such larger simulation dumps is a time-consuming process. Thus, we assume that $t_s \gg t_p$.

Comparing (8.5) and (8.6), overall power analysis speed up obtained by using the proposed methodology can be given as

$$N_{PE} = n_r/n_s. \tag{8.7}$$

Applying (8.4) in (8.7), we get

$$N_{PE} = \beta + (n_i/n_s). \tag{8.8}$$

Value of β depends on the TLM specification style, for most of the designs $n_i > n_s$ which also contributes to the overall speedup offered by the proposed power analysis methodology with respect to RTL power analysis approach. Our experiments as shown in Table 8.1 validate the above arguments. Certain assumptions used in this analysis may not hold in the practical scenario; however, they are used here to prove the applicability of our approach.

This proposed approach for power analysis can be called as mixed probabilistic approach. During this process, activities of the input–output ports and some intermediate signals are correctly imported from the high-level simulation dumps. Using this information, faster probabilistic power estimation is done at RTL which contributes to the overall speedup. Based on the accuracy of probabilistic power estimation, overall results can be improved. Furthermore, an accurate mapping of the high-level variables to the corresponding RTL signals also contributes to the accuracy of such a power analysis methodology. Some high-level synthesis tools generate such a mapping file which can be used to further improve the power estimation accuracy of our approach.

8.3 Our Methodology

Our high-level power-estimation methodology starts with the high-level model of a hardware design and generates the corresponding power numbers. As shown in Fig. 8.1, the overall methodology can be divided into the following steps:

1. Convert the high-level model of the design to an equivalent cycle-accurate RTL model. This can be done manually or using an appropriate high-level synthesis engine. Most high-level synthesis tools [28, 101] facilitate capturing the specification of a design at a level of abstraction above RTL followed by automatic generation of the corresponding RTL code. In our experiments, we have used Esterel Studio to synthesize the high-level ESTEREL models to their RTL implementations.

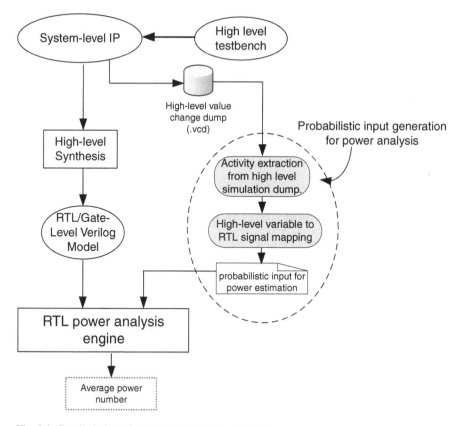

Fig. 8.1 Detailed view of our power estimation methodology

2. Simulate the high-level model and generate its VCD. For better accuracy of power numbers, all high-level variables of the model should be captured in the VCD.
3. Apply Algorithm *VCDExtraction* (Sect. 8.3.1) on the VCD file generated during high-level simulation in Step 2. Algorithm *VCDExtraction* is used for activity extraction from the high-level simulation dump.
4. Generate the mapping of high-level variables to RTL signals as described in Sect. 8.3.2 and use the outputs of Algorithm *VCDExtraction* to create appropriate inputs for performing probabilistic power analysis using RTL power estimator (such as *PowerTheater* [129]). Such inputs include power-related information for the mapped RTL inputs, outputs, and other intermediate signals, which can be then probabilistically propagated by the RTL power estimator for finding the activity information of the remaining signals and generate the power numbers.

In our framework, high-level power-related information extracted from the algorithm *VCDExtraction* is provided as input to PowerTheater in a particular format known as global activity format (GAF). Note that in pure RTL

probabilistic power-estimation technique, such inputs (information related to the activity of ports and some intermediate signals) are usually provided by the user, and in some cases default values of activity factors are used. On the other hand, for the mapped variables/signals of a design that are visible at both high-level as well as RTL, their high-level simulation information extracted in our framework is exactly same as the information which would have been generated during the RTL simulation, and hence our approach can provide better results than pure probabilistic approach.
5. Collect the power numbers reported by the RTL power estimation tool.

8.3.1   Activity Extraction from the High-Level VCD

Activity and duty-cycle extraction of various variables of a design is an integral part of our proposed methodology. Consider a high-level design S with the following notations:

$M = \{m_i \; : \; m_i$ is a module of design $S \}$; $V_i = \{v_{ij} \; : \; v_{ij}$ is a variable used in module $m_i \in M \}$.

Algorithm *VCDExtraction* extracts the hierarchy, activity-factor and duty-cycle related information from a VCD file generated during high-level simulation. Activity-factor and duty-cycle are computed using, (8.9) and (8.10) respectively.

$$\text{Activity factor,} \, a_k^{ij} = (N_k/N_{\text{h}}) \tag{8.9}$$

$$\text{Dutycycle,} \, \delta_k^{ij} = (t_k/t_{\text{total}}) \tag{8.10}$$

In (8.9), N_k denotes the total number of times a variable's value changed during the simulation and N_{h} represents highest toggles that can occur on a variable's value (in general, clock of the design has the highest toggles). In (8.10), t_k represents the time duration for which the kth bit of a variable remains in a boolean value *True* and t_{total} represents the total simulation duration.

Input to the Algorithm *VCDExtraction* is a high-level VCD file and its output is the duty-cycle, activity-factor and hierarchy information associated with each signal of the RTL design. The algorithm can be divided into two parts:

1. *Part a* Extraction of following information: (1) hierarchy for each variable, (2) for each value of a variable, number of simulation ticks for which the value remains unchanged, and (3) order in which various values of a variable changed during the simulation.
2. *Part b* Calculate the activity-factor and duty-cycle of various signals using (8.9) and (8.10), respectively.

Algorithm *VCDExtraction*
Input : High-level VCD file.
Output: Hierarchy, Activity and Duty-cycle of various RTL signals.

Procedure *Part (a)* {
$\forall\ m_i \in M$ **do**
   $\forall\ v_{ij} \in V_i$ **do**
      a.1- Store v_{ij}'s hierarchical name w.r.t the topmost module.
      a.2- Collect the changes occurred in the value of v_{ij}
            and the order in which its values changed.
      a.3- Estimate the simulation time spent in each value.

}
Procedure *Part (b)* {
$\forall$ bit b_{ijk} of v_{ij} from LSB to MSB **do**
   b.2- Using (8.10), and information estimated
         in *Part a.3*, estimate the duty-cycle of b_{ijk}.
   b.1- Using (8.9) and information collected
         in *Part a.2*, calculate the activity factor of b_{ijk}.
}

Return hierarchical information, duty-cycle and activity-factor for various signals
of the design.

8.3.2   High-Level Variable to RTL Signal Mapping

High-level model of a design is an abstract model capturing the functional behavior
of the design, and hence it contains much lesser information compared to the
corresponding RTL or gate-level implementations. An ESL designer's job is to make
sure that high-level model is compliant to the specification of the given design and
to ensure, via architectural trade-offs, that the implementations generated (using a
high-level synthesis tool) from such a model meets all the constraints on the area,
power and latency of the design. Our proposed power-estimation methodology can
be used in such scenarios. When synthesis tools create RTL models of designs
from their high-level models, they provide either a mapping file containing the
information about high-level variables and corresponding lower-level signals or
some guidelines to find such relationships between the high-level variables and
lower-level signals. Esterel Studio provides such a facility for mapping inputs,
outputs and intermediate variables of the high-level model to the RTL signals. In this
methodology, we used this mapping to map activity information of all the high-level
variables to RTL signals for power estimation purpose.

Variables at high-level may or may not have same name in the generated RTL code, and thus proper mapping of high-level variables to RTL signals should be created for accurate power estimation. We briefly provide an overview of the information we extracted from our experimental setup. In this paper, we extracted this information based on certain rules or synthesis guidelines which are specific to Esterel Studio synthesis. More detailed information exists in the user manual of Esterel Studio [60]. It should be noted that generating such a mapping is one time process for the analysis of a design because once this information is available then it can be used for many simulation runs. Activity of high-level variables extracted from algorithm *VCDExtraction* is then used for the generated signals/nets (in case their name gets transformed because of HLS).

8.3.2.1 Specifics for Our Experiments

We briefly provide overview of the information we extracted for our experimental setup. Note that different synthesis tools may provide such information in different formats, and for our experiments, we extracted this information based on certain rules or synthesis guidelines which are specific to Esterel Studio synthesis. More detailed information exists in *monolithic HDL code generation chapter in user manual of Esterel Studio* [60]. Below, we capture the salient points considered for our analysis.

- A signal that is exchanged between a module and its external environment is called an interface signal (similar to input output in HDLs). Once the RTL code is generated the interface signals are flattened in generated code. Listing 8.1 shows an example:

Listing 8.1 Interface in ESTEREL

```
interface A:
  input i;
  output o;
end interface
interface B:
  input j;
  port p: A;
end interface
.....
port q: B
```

In the example shown in Listing 8.1, B has a port p of type A and port q is defined as of type B. While generating the RTL code for port q, Esterel Studio will flatten the input and output variables and in generated RTL code these variables looks as shown in Listing 8.2:

Listing 8.2 Interface signals in generated Verilog

```
input  q_j;
input  q_p_i;
output  q_p_o;
```

- Signal names such as $V7\_sb\_X$ may appear for intermediate signals used for connecting logic-gates, etc. in the generated RTL corresponding to the high-level variable X.
- A module M having an interface or intermediate variable X at high-level will be translated into two signals. Suppose the variable is represented as:

input X : bool;

then the generated RTL code will see the following signals:

input X;
input X_data;.

ESTEREL has the notion of full signals; generated RTL code will see two signals corresponding to the status (X) and actual value (X_data), where status is an extra control information, if status is *true* then actual value will be updated else it will remain unaffected.

One should note that generating such a mapping is one time process for the analysis of a design because once this information is available then it can be used for many simulation runs. Activity of high-level variables extracted from algorithm *VCDExtraction* is then used for the generated signals/nets (in case their name gets transformed because of HLS).

8.4 Results

Tools Used: In our experimental setup, we have performed RTL and high-level simulation using VCS [149] and Esterel Studio [60] respectively. We have used 180 nm power characterized *typical case* process libraries. We have used VCS 2006.06 release for RTL simulation, Esterel Studio 5.4.4 (build i2), and PowerTheater 2008.1 release for power estimation purposes. We have performed all our experiments on Dell optiplex (GX620) machine (operating system redhat enterprise linux (RHEL)) with 2 GB RAM and 3.39 GHz CPU speed.

Experiments: Table 8.1 captures the results obtained on different design IPs modeled at high-level. Entries in Column 1 (Design) represents the design type, column 2 (P_{RTL}) represents the power numbers obtained using RTL power estimation technique, Column 3 (P_{SPAF}) represents the power numbers obtained using our methodology and column 4 (E) shows the error with respect to the RTL power estimation approach. Column 5 (G_{VCD}) represents the ratio of the size of VCDs obtained at RTL and high-level to provide an overview of the difference of information needed at lower-level design flow.

Among various design IPs, we have used VeSPA processor model (which is a 32-bit processor with five stage pipeline) [98], and implemented the design in behavioral style, for this case study model is written as instruction set simulation model in ESTEREL. Other examples in this case study include power state machine

Table 8.1 Results of our approach on different designs

Design	P_{RTL} (mW)	P_{SPAF} (mW)	E (%)	G_VCD
PSM	1.23	1.16	5.7	168
FFT	12.6	13.7	8.7	50
FIR	21.5	22.2	3.3	1,189
RAM	0.451	0.419	7.12	8
VeSPA	132.4	125.95	4.87	60
UART	11.7	10.8	7.69	33

Table 8.2 VeSPA processor instruction wise power numbers

Instruction Type	P_{RTL} (mW)	P_{SPAF} (mW)	E (%)
ADD	159	153	3.77
AND	119	107	10.08
NOT	130	131	0.77
XOR	151	150	0.66
OR	125	123	1.6
CMP	159	153	3.77

(PSM), which is a prototype of a controller/state-machine used in modern processors for power management purposes. Other designs are finite impulse response filter (FIR), fast Fourier transform (FFT), universal asynchronous receiver and transmitter (UART) etc.

$$E = |(P_{RTL} - P_{SPAF})/P_{RTL}| * 100 \tag{8.11}$$

$$G_{VCD} = G_{RTL\_VCD}/G_{SPAF\_VCD} \tag{8.12}$$

In (8.11), P_{RTL} is the power number obtained using traditional RTL power estimation approach and P_{SPAF} is the power number obtained using our approach. In (8.12), G_{RTL_VCD} represents the disk space used by VCD in RTL power estimation and G_{SPAF_VCD} using (8.3).

We can see difference in accuracy (Tables 8.1 and 8.2) for various benchmarks is ranging from 3–9% We attribute these differences to various facts (a) these models are modeled by four different people so modeling style might be different among the modelers for the same synthesis guidelines; (b) probabilistic activity propagation may suit particular type of logic/design, e.g. most of the instructions in VeSPA processors except AND instructions for random test patterns are showing good accuracy (within 4%); (c) there can be scenarios where intermediate signals at RTL simulation (that cannot be related with the high-level model) have very high activity factor but in our methodology power estimation tool is propagating the activity value based on fixed probability. But one should note that results are consistently within 10% and can be further improved if one can establish better ways of establishing activity relationship between generated intermediate RTL signals and high-level variables.

Earlier presented work for the high-level models (such as [34, 113]) also provide the similar results. Approach discussed in these references utilize the power models for power estimation (mostly in specific scenario or for particular application). However our approach is an automation solution, which aids high-level synthesis framework to get RTL like accuracy with some extra effort.

Power Estimation Speed-up: We have seen 2–12 times speedup for our benchmarks. Power estimation time speedup for FIR design using our approach is 12 times and speedup in activity extraction time t_{s_a} is 26 times. We compared the speedup numbers by running the designs for few minutes of CPU time to obtain these speedup numbers. In case of VeSPA design where we got the simulation bug, we ran few seconds simulation to get the accuracy numbers (random tests of 10,000 clock cycle duration). We found that for smaller simulations tool takes almost constant time in seconds, which can be attributed to the fact that tool require some initial setup time to run these simulations, hence bigger runs are required to compare the speedup numbers. Such bigger simulations caused a simulation bug (out of memory) at high-level and we could not report the speedup numbers for all the designs. Efficiency of our approach is very much dependent on the amount of information being processed by the power estimator. As shown in the Table 8.1, upto 1,100 times disk space reduction is possible for simulation dump size.

Chapter 9
Applying Verification Collaterals for Accurate Power Estimation

9.1 Introduction

In this chapter, we present a methodology for accurate power estimation at high-level, which utilizes the existing verification or validation resources in the design flow. This methodology can help in developing an infrastructure to estimate the power consumption at higher-level. Current power estimation methods are limited to RTL or lower level, in our approach we utilize best from both the worlds (RTL and high-level) and try to get the RTL like accuracy using high-level simulation only. We provide steps, examples of different properties using assertions, and methods to create directed testbenches as verification resources in the design flow for power estimation purpose. The approach presented in this chapter is specific to a high-level synthesis framework. We use Esterel Studio and Power Theater (to show the efficacy of our results). Our case study is performed on a finite state machine (FSM) based design. This case study presents detailed description of properties to generate the directed testbench for reaching a particular state of the FSM. Once we have the stimulus for the particular state of the FSM, we estimate the power of that particular state of the design at RTL, which is further utilized for the high-level power estimation.

Our approach utilizes verification/validation resources existing in the design flow to activate different key states and capture power related information. Simulators, checkers, coverage models, assertions, test content, and tools developed during a typical validation flow are collectively called *verification collaterals*. Verification collaterals are employed to enhance correctness and to improve confidence in the high-level model that it satisfies the design requirements. Our focus in this chapter is specific to the creation of directed testbenches or the development of assertions and how they can be utilized downstream to characterize power numbers for different aspects of a high-level design such as states and transitions. The advantage of our approach is the ability to: (1) reuse existing verification collaterals and (2) estimate power at high-level without rebuilding the complete infrastructure. Such an approach helps in quickly narrowing down power-specifics earlier in the design cycle, which facilitates a realistic and rapid power exploration for a design.

S. Ahuja et al., *Low Power Design with High-Level Power Estimation and Power-Aware Synthesis*, DOI 10.1007/978-1-4614-0872-7_9,
© Springer Science+Business Media, LLC 2012

9.2   Our Methodology

Figure 9.1 shows the detailed view of our technique at system-level utilizing the average power associated with each mode and transition. Figure 9.2 provides the detail on how to obtain the average power number associated with each mode and transition. For applying our power estimation technique, designers require a high-level designing and verification framework. We are using Esterel Studio for designing and verifying the High-level models (HLM). Our HLMs are written in cycle-accurate transaction level (CATL) style in ESTEREL. From high-level simulation (performed in Esterel Studio) we extract the total time spent in each state, transitions and the total number of transitions. We also utilize the average power consumption in each mode at RTL (using PowerTheater) and calculate the power at system-level. The main advantage of our approach is that we do not require to do the power estimation at RTL for every testbench from the testsuite for the whole design. We utilize average power associated with each mode for estimation purpose at system-level.

Following are the steps that needs to be followed for power estimation in this framework (Fig. 9.1):

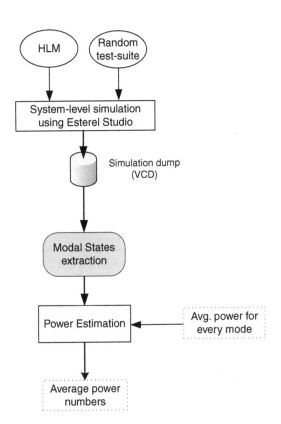

Fig. 9.1 Detailed view of system-level power estimation technique

Fig. 9.2 Power estimation
for each mode at RTL

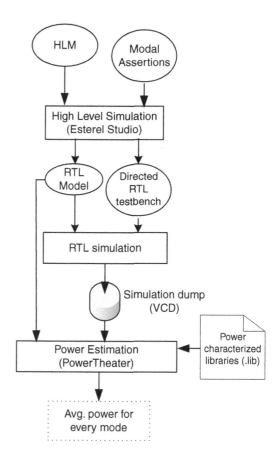

1. Simulate the High Level Model (HLM) written in ESTEREL using Esterel Studio and store the Value Change Dump (VCD) for further analysis.
2. Extract different modes and transitions, total number of transitions, total time spent in each mode, and transitions from the simulation dump as discussed in Sect. 9.2.2.
3. Once we have the estimate on time spent in each state and transition, power estimate of each state, and transition then calculate the average power of the design using system-level simulation as discussed in Sect. 9.2.3. This utilizes the average power number for different modes using assertions as shown in Fig. 9.2. Here are the steps to obtain the average power number for each mode/transition (Fig. 9.2):

    a. We create the cycle-accurate RTL model corresponding to the system-level cycle-accurate description in ESTEREL using the high-level synthesis engine provided by Esterel Studio (ES). ES is also used to convert the modal assertions (as discussed in Sect. 9.2.1) to directed RTL testbenches.

b. Once the RTL model and testbenchs are ready, we simulate the RTL model using RTL simulator (VCS [149]) and generate the simulation dump.
c. The dump generated at RTL, power characterized libraries and all the required inputs are then passed to the RTL power estimation tool (PowerTheater). The average power numbers calculated at this stage are then utilized for power estimation during system-level simulation. This is one time process for as these numbers can be utilized for multiple system-level simulations for power estimation.

9.2.1 Assertions for Finding out Particular Mode of Design

In a typical design flow, properties are written for verifying the behavior of the design. In a design flow where design can work in different modes, one can write assertions to excite the different modes (states) of a design and to trigger a mode change (transition). We assume that the properties are expressed as assertions and enforced during the verification stage of the design. Such that they are available during the power estimation stage, if not then we write the assertions for testing the reachability of states and transitions. We illustrate some reachability assertions written for a simplistic modal design shown in Listing 9.1. Note that the design is specified as a pseudo-code and not in any specific language.

Listing 9.1 A simple modal design

```
module XYZ
begin
  bool c1, c2;
  string state;    // Possible values A, B, C, D

  if (c1 == 0 && c2 == 0) then
    state = "A";
  else
    if (c1 == 1 && c2 == 0) then
      state = "B";
    else
      if (c1 == 0 && c2 == 1) then
        state = "C"
      else
        state = "D"
      end if
    end if
  end if
end
```

The design shown in Listing 9.1 has a variable state that is used to capture the mode of the system, which basically has four different values. The Boolean control variables c1 and c2 are used to trigger the required mode change. Assertions written for state reachability properties are shown in Listing 9.2.

Listing 9.2 Assertions to verify the reachability of states

```
// Property specifies the condition that reaches state A

assert always ( (c1=0 && c2=0) -> (state='A') );

// Property specifies the condition that reaches state D

assert always ( (c1=1 && c2=1) -> (state='D') );
```

An assertion that triggers the system to change from one state to another is shown in Listing 9.3.

Listing 9.3 Assertions to verify the reachability of a transition

```
// Property specifies the conditions that causes a transition from state D to
   C

assert always ( (c1=1 && c2=1 && next (c1=0)) -> ((state='D') && next(state='
   C')) );
```

The assertions are utilized for creating directed test cases such that each testbench either drives the system to a mode or causes the system to change its mode of operation. We utilize Esterel to express the assertions and Esterel Studio to generate the corresponding testbenches. Note that these directed testbenches are given as input to PowerTheater to estimate the expected power in the various operating modes of the system. These power numbers are reused at system-level to compute the overall power consumption of the system.

9.2.2 Extraction of Modes from the Simulation Dump

As discussed earlier at high-level there are control signals associated with each of the mode, any change in the value of these control signals is dumped in the simulation-dump. Signals for these modes can have value depending upon its type, e.g. if there is a boolean associated with mode then we check when the signal is true, else if it is short then we check the exact value of the variable associated with that mode. We extract all the possible time-stamps for which the mode signal remains in the expected value and how many number of times design goes to particular mode for calculating the total time spent in each mode. This can easily be done as VCD contains all the information related to the value of the mode. Similarly, we calculate the total number of transitions (from one mode to another) occurring in the dump. This knowledge is then used in doing system-level power estimation of the design. We wrote a C-shell script to extract the total time spent in each mode/transition, and the number of times design comes to that mode/transition.

9.2.3 High-Level Power Estimation

To calculate the high level power we first try to establish a relationship using energy spent in each mode during the full simulation duration and then we establish a relationship for the power at each state. Lets say there are n different states and m different transitions in the design. Energy spent for the state i can be represented as E_i and total time spent in the state i can be represented as t_i. Similarly, E_j is the energy associated with transition j, t_j is the total time spent on the transition j. If E_{total} is the total energy consumed in the design then we can establish the following:

$$E_{total} = \sum_{i=1}^{n} E_i + \sum_{j=1}^{m} E_j \tag{9.1}$$

If total average power is P_{total} and total simulation duration is T, then from (9.1) we can establish the following:

$$P_{total} * T = \sum_{i=1}^{n} P_i * t_i + \sum_{j=1}^{m} P_j * t_j \tag{9.2}$$

If a_i, a_j is represented as a fraction of total simulation duration spent in state i and transition j, we establish a relationship between a_i, a_j, t_i, t_j and T as shown in (9.3) and (9.4).

$$a_i = t_i/T \tag{9.3}$$
$$a_j = t_j/T \tag{9.4}$$

In (9.2), if we divide both the sides by T then we will get the following:

$$P_{total} = \sum_{i=1}^{n} P_i * a_i + \sum_{j=1}^{m} P_j * a_j \tag{9.5}$$

which can further be simplified as

$$P_{total} = \sum_{i=1}^{n} P_{Hi} + \sum_{j=1}^{m} P_{Hj} \tag{9.6}$$

In (9.6), P_{Hi} and P_{Hj} represents the component of expected average power spent in each state and transition, respectively. Equation, (9.5) and (9.6) establishes the relationship between average power of each state, transition and total power calculated at high-level. In our approach, P_i and P_j is calculated at RTL by utilizing the assertions written during verification stage (as discussed in Sect. 9.2.1) and a_i and a_j is calculated as discussed in Sect. 9.2.2. Finally, P_{Hi}, P_{Hj} is calculated from the values of P_i, P_j, a_i, and a_j.

9.3 Case Study

9.3.1 PSM Specification

State diagram of the PSM is shown in Fig. 9.3, more details about the design is available at [118].

Fig. 9.3 Macro-state
diagram of PSM

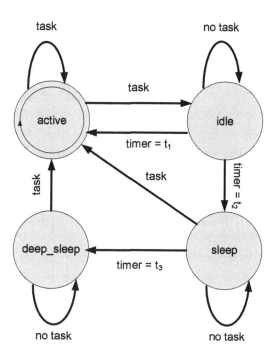

PSM is an essential ingredient in reducing the power consumption of the system by
regulating the power states. It is distributed along the design; hence, the validation of
the PSM requires simulating almost the whole system. In Fig. 9.3, we show the four
different states of the PSM namely active, idle, sleep, and deep_sleep. Its
four different constituents namely *Queue*, *Timer*, *Sampler*, and *Service Provider* are
shown in Fig. 9.4.

- *PSM:* The system remains in active state if it is receiving a task. If no task is
  received for t_1 clock cycles, then it goes to the idle state. In the idle state, if it
  does not receive any task for t_2 cycles, then it goes to the sleep state. In sleep,
  if the system does not receive any task for t_3 cycles, then it goes to the deep sleep
  state deep_sleep. From any of these states, the system returns to the active
  state if it receives a task.
- *Sampler:* All power states have a sampler that samples a task every t_4 cycles.
- *Queue:* A task received by the system is stored on a finite queue, which is
  accessible to every state.
- *Service Provider (SP):* In the active state, the SP processes a sampled task for
  t_5 cycles. The SP requests for a task and receives a task that is de-queued.
- *Timer:* Every state except the deep_sleep has a timer embedded, which tracks
  for how long the system is idle without receiving a task.

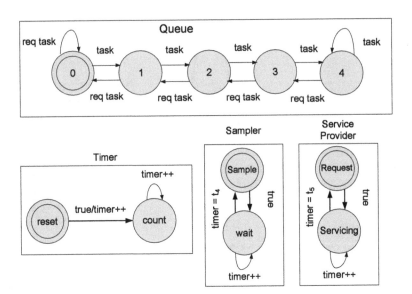

Fig. 9.4 Specification of the different PSM ingredients

Verification of the PSM is very important because such a component is often used in system level power management for reduction of power in embedded systems and this type of controller fits the design type we are targeting in this chapter.

One can reach to a particular state by putting the property in form of assertion on the state. Same stimulus can be created manually by writing the directed testbench from the specification of the design. We explain both the approaches in detail in this section.

9.3.2 Directed Testbench Creation from the Specification

We illustrate this approach by the example specification discussed in the Sect. 9.3.1. Let us assume that we want to obtain the idle state power numbers of the design. As discussed in the Sect. 9.3.1, design goes to idle state (from active state) if it does not receive any input task for t_1 cycles. Design will remain in the idle state for t_2 cycles if it still does not receive any task. Hence, one can write a testcase in which design will not receive any input task for t_1 cycles, after that design will remain in the idle state from t_1+1 until $t_1 + t_2$ cycles. To capture the power information of the idle state one should run power estimator from (t_1+1) to $(t_1 + t_2)$ clock-cycles to capture the average power consumption of the design in idle state, although stimulus should atleast have continuous $(t_1 + t_2)$ clock-cycles on which design does not receive any input task.

Similarly, one can write the testcases for all the valid transitions. For example, in the case discussed above at cycle t_1, the transition from active state to idle state will occur. If, there is a good power management system working in the system then lot of computation unit in the design won't be in active state and hence in such a situation there will be sudden drop in the power consumption of the design. Now, to obtain transition power from one state to another instead of measuring the average power over certain clock cycles, we did clock by clock power measurement for the clock cycles when these transitions are occurring and collected the power consumption for the transitions. Such measurements help us in improving the accuracy of the average power numbers.

We can observe there are four different macro states, i.e. active, idle, sleep, and deep_sleep state, respectively. From the state transition diagram, we see that one can write testcases for all the valid transition scenarios that include active to idle, idle to sleep, sleep to deep_sleep, idle to active, sleep to active, and deep_sleep to active, etc.

The detailed explanation on how one can obtain the power estimation at high-level is provided in [10]. In this section, we briefly provide the description on how we have used that method in our case. As discussed above, we can first create directed testbenches from the specification and obtain the average power numbers for these scenarios using RTL power estimator PowerTheater [129].

We generate the high-level simulation dump (VCD), which has the information of these state variables, for any arbitrary testvector. We calculate the time spent on each state during the simulation, from the VCD. We then calculate the total energy spent on each state, which is the product of power and time spent, i.e. clock-cycles in a particular state, with the information extracted from the earlier steps. Then we sum up the energy for each state/transition and average it over whole simulation duration to calculate the average power.

9.3.3 Utilizing Assertions for Creating Directed Testbench

Various properties to reach/stay in a particular macro-state can be written for the design. The purpose of this step is to provide an approach to automate the testbench generation process using simple assertions. We propose two type of approaches: (a) By writing a verification monitor for the whole design from which one can monitor input and output values and then insert a property to see weather design follows it. (b) As a part of design, writing some properties on states or timer values to see if design can ever reach to those states or values.

9.3.3.1 Writing Properties Using Verification Monitor

For verification of certain properties, we create a new toplevel in which we put the design as *design under verification (DUV)*. From the verification monitor, we can control the inputs, observe the outputs, and verify the properties of the design.

- *Assertion to reach a particular state of the design*: As explained earlier to reach to idle state of the design, it should not receive any input task for t_1 cycles and if we want the design to stay in this particular state then input task should be kept false for $(t_1 + t_2)$ clock-cycles. Listing 9.4 shows how one can write the assertion whose negation will force the Esterel Studio to provide a counter case that design will go to idle state. In the listing shown, we force input task not to receive any input for $(t_1 + t_2)$ clock-cycles using repeat construct. One can use the for loop, etc. Just after that we put an assertion saying that input task can never be true. Note that we set this property as an always true property and then ask Esterel Studio to prove this property, for which the verification engine generates a testcase showing that input can be false for more than $(t_1 + t_2)$ clock-cycles. State specific power estimation in this case will be similar to the one explained in Sect. 9.3.2.

Listing 9.4 Assertion for state reachability

```
repeat ?t1+?t2 times
  emit ?input_task = false;
  await tick;
end repeat;
assert idleStateReachability = not(?input_task = false);
```

- *Assertion for a transition from a state*: Example explained above will contain the transition from active to idle state because no task was received till $(t_1 + t_2)$ clock-cycles. Similarly, idle to sleep and sleep to deep_sleep transitions can be captured using the property shown above. For transitions from idle or sleep or deep_sleep state to active state transition cannot be generated using the property above. Listing 9.5 shows the transitions scenario that can be captured using Esterel Studio which is written as never true property. Now the never true property is written in such a way that we force the verification monitor not to receive any input task for certain clock-cycles and then we ask that design cannot receive any task further. The counter case of this case is putting no input task till certain cycles and then allow the input task, which forces the design to move to active state from idle/sleep/deep_sleep state, etc. Listing 9.5 shows the transition to active state from the deep_sleep state.

Listing 9.5 Assertion for state transition

```
repeat ?t1+?t2+?t3 times
  emit ?input_task = false;
  await tick;
end repeat;
assert idleActiveTransitionStateR = not(?input_task = true);
```

In this case study, we show controlling one input because that can help us in identifying the state but this approach can specially be helpful for the cases where user wants to control the multiple inputs in parallel. In that case, such an approach can be used to automatically generate testcase using assertions.

9.3.3.2   Writing Properties on Design States/Variables

- *Generating a testcase to reach a state*:
  As discussed earlier, one can write assertions in the design and can set the
  property as always true or assumed using Esterel Studio. In case the property
  is not true, then Esterel Studio generates a counter case for the scenario. For
  PSM, the code snippet for the design from high-level utilizing the macro-state
  behavior looks as shown in Listing 9.6.

Listing 9.6   Overview of the code for states in Esterel

```
if (? State = "active") then
.........
.........
else
    if (? state = "idle") then
    .........
    .........
    end if;
else
    if (? state = "sleep") then
    .........
    .........
    end if;
else
    if (? state = "deep_sleep") then
    .........
    .........
    end if;
end if;
```

Now, to create a testbench to reach to a particular state, one can write an assertion
inside the design code of that state and putting a property that design can never
remain in that state itself. For example, if we want to see if design can reach to
the idle state of the design then inside the code for the idle state, we need to write
an always true property, which says that in idle state, design will never remain in
the idle state (which is a contradiction). So the code for such a scenario will look
as shown in Listing 9.7.

Listing 9.7   Assertion for state reachability

```
if (? state = "idle") then
.........
.........
assert idle_state_check = not(? state = "idle");
end if;
```

For the scenario stated above Esterel Studio generate a counter example that
contains input which forces the design to remain in the idle state.

- *Testcase to reach a particular state and then remaining inside the state for some
  cycles using invariant*:
  We have seen in the specification section that design remains in particular state
  for some cycles. We can write properties, in which we can verify, if design
  remains in that state for some time or not. For PSM example one can do the same
  by using the invariants for the timer. In our high-level design, we have a separate
  timer module. In Esterel studio, we can write an assertion shown in Listing 9.8.

Listing 9.8 Assertion for staying in a state for certain cycles

```
assert idle_timer_check = not(? timer_count = ?t2);
```

Now for the case shown in Listing 9.8, we can set the Esterel Studio to test an always true property for the instantiation of the timer for idle state. From the specification document, we know that timer will always go to t_2, if it does not receive any input task. Esterel Studio will be able to produce a counter example for the negated property.

- *Testcase for generating Transition Scenario*:
  One can write the assertions to generate testcases in which design goes through the state transitions. Listing 9.9 shows such an example of the transition condition that can be used for creating the directed testbench which can be used to capture the macro state transitions.

Listing 9.9 assertion for extracting transitions

```
assert state_test = not ( (? state = "active") and (next ? state = "idle"));
```

Assertion shown in the Listing 9.9 represents the valid transition extracted from the specification document. Now this assertion is again negated and Esterel Studio option for always true is set to test such a property. From specification, we know that such case is trivial transition case hence Esterel Studio generates a negated scenario that can be used to capture the state transition of the design. The Listing 9.9 shows that design can never reach to idle state from active state, which is not true. In this case, Esterel Studio generates a counter case showing that design can reach to idle state from active state.

9.4   Results

In this chapter, we presented a case study of our approach on power state machine. We presented various properties one can write to obtain state specific typical stimulus input. We also presented an approach in which one can write those tests manually based on the specification provided for the design. Our results utilizing the manual approach shows good results with good accuracy. We automated the manual testbench creation using the approach discussed in this chapter. We presented our case study on small design, while the scalability of such an approach still needs to be tested on bigger designs. For example, using invariant based approach might be very costly sometimes, if design has a lot of states. Similarly, capturing transitions for different states may take many cycles in that case, we will require instrumentation of the source code according to that specification.

As discussed in the Sect. 9.2.3, we calculate the time spent in each state, transition (a_j, a_k) for the whole simulation and the results are captured in the Tables 9.1 and 9.2. P_j and P_k in (9.5) represent average power of each state and transition captured in Tables 9.3 and 9.4. P_{Hj} and P_{Hk} in (9.6) represents the weighted per state power and transition power, these numbers are reported in Tables 9.5 and 9.6. We calculated the total average power using our methodology

Table 9.1 Percentage of simulation time spent in each state at system-level

PSM state	Active	Idle	Sleep	Deep_sleep
% of total-time spend in each state	56.68	10.44	19.63	13.23

Table 9.2 Percentage of simulation time spent in each transition

Transitions	% of total-time spent in each transition
Active→idle	0.013
Idle→sleep	0.013
Sleep→deep_sleep	0.007
Deep_sleep→active	0.007
Sleep→active	0.007
Idle→active	0.000

Table 9.3 Power estimation of the states at RTL

State	Active	Idle	Sleep	Deep_sleep
Power consumption (mW)	1.57	0.834	0.833	0.833

Table 9.4 Power estimation at RTL for each transition

Transitions	Power consumption (mW)
Active→idle	1.61
Idle→sleep	1.61
Sleep→deep_sleep	1.62
Deep_sleep→active	12.72
Sleep→active	12.72
Idle→active	12.72

Table 9.5 Power numbers at each state using our approach

State	Active	Idle	Sleep	Deep_sleep
Power consumption (mW)	0.8898	0.0871	0.1635	0.1103

Table 9.6 Percentage of simulation time spent in each transition

Transitions	Power consumption (mW)
Active→idle	0.00021
Idle→sleep	0.00021
Sleep→deep_sleep	0.00011
Deep_sleep→active	0.00089
Sleep→active	0.00089
Idle→active	0.00000

Table 9.7 Estimation accuracy as compared to RTL

RTL average power (mW)	System-level average power (mW)	Accuracy w.r.t RTL (%)
1.270	1.253	98.66

applied at system-level simulation and got the power number 1.253 mW for a randomly selected testbenches. To verify our result we created the equivalent RTL testbenches of the design for which we calculated the power using system-level testbenches and found out the power number P_{RTL} as 1.27 mW. Accuracy (E) of our power estimation methodology in this case is coming out as 98.66%, E is calculated using (9.7). All these results are captured in Table 9.7.

$$E = (P_{total})/P_{RTL}) \times 100 \tag{9.7}$$

From the result tables we can see that the minimum and maximum bound on power for this design ranges from 0.833 mW to 1.57 mW. We can also notice that if design has got a lot of transitions from idle, sleep, deep_sleep to active state, or vice versa then we will also see a lot of variation in average power of the whole design because these transitions consume a lot of power Transitions in this case are taking a lot of power, which is also understandable because if the system is coming from sleep state to active state then most of the design will see a lot of activity and hence a huge amount of power. Overall effect on total power is not too high because of transitions. Total time spent in transitions is less than 1% as shown in the Table 9.2.

Chapter 10
Power Reduction Using High-Level Clock-Gating

Hardware co-processors are used for accelerating specific compute-intensive tasks dedicated to video/audio codec, encryption/decryption, etc. Since many of these data-processing tasks already have efficient software algorithms, one could reuse those to synthesize the co-processor IPs. However, such software algorithms are usually sequential and written in C/C++. High-level Synthesis (HLS) helps in converting software implementation to RTL hardware design. Such co-processor based systems show enhanced performance but often have greater power/energy consumption. Therefore, the automated synthesis of such accelerator IPs must be power-aware. Downstream power savings features such as clock-gating are unknown during HLS. Designer is forced to take such power-aware decisions only after logic synthesis stage, causing an increase in design time and effort. In this chapter, we present a design automation solution to facilitate various granularities of clock-gating at high-level C description of the design.

10.1 Introduction

Adoption of Electronic System Level (ESL) based methodologies in hardware design flow is on the rise. C/C++/SystemC are the description languages used to represent the design behavior at the ESL. One of the main ingredients of ESL based design flow is high-level synthesis (HLS). HLS helps in creating RTL/gate-level description of the design from the behavioral or system level description. In the recent past, use of hardware co-processors was advocated for specific domains such as digital signal processing (matrix multiplication, filtering operations), security (encryption/decryption), etc. These signal processing or security related tasks are focused on data-intensive computing for which efficient software algorithms exist in C/C++. Earlier these tasks were exclusively done in software. With the availability of more gates on the silicon die, use of co-processor/accelerator to perform these data-intensive computations using co-design methodologies is on the rise.

S. Ahuja et al., *Low Power Design with High-Level Power Estimation and Power-Aware Synthesis*, DOI 10.1007/978-1-4614-0872-7_10, © Springer Science+Business Media, LLC 2012

While such co-processor based acceleration can enhance performance, it might impact power/energy consumption of the system. Therefore, the automated synthesis of such accelerators must be power-aware.

HLS of such efficient software algorithms will greatly help in reducing design cycle time as well. It is beneficial to reuse these time-tested and validated software algorithms. It is even more beneficial if one can use automated synthesis to convert these co-processor IPs from their software implementations, rather than manually recasting the algorithm into low level hardware implementation. One hurdle in such a strategy is that these software algorithms are targeted usually for sequential processing, written in C/C++. In contrast, the hardware implementation has to be highly parallel, clocked, leading to both efficiency and low power requirements. The hardware acceleration require intensive parallelism/pipelining/buffering techniques which mandates appropriate choice of their micro-architectures (parallel, pipelined, etc.).

The process of making this choice may require multiple passes through refinement/synthesis followed by performance/power estimation cycles. In a power-aware methodology, the power estimation must be accurate, and at the highest possible level of abstraction for faster convergence of the selection process. However, down-stream gate-level power savings features unseen at register transfer level (RTL), may render the exaggeration of the power estimates and force the designer to take decisions only after logic synthesis and estimation. This increases design time and effort. In this chapter, we show how to endow the HLS itself with the ability to generate RTL with the power-saving features that are normally inserted during gate-level synthesis. This allows realistic power exploration at the RTL, resulting in faster convergence of the micro-architecture selection, without compromising the quality of the generated hardware co-processor.

In an HLS based design flow, tool generated RTL model is passed through logic synthesis tools that would add various power saving features such as clock-gating. These power reduction features are often unseen at high-level. Therefore, the RTL power estimation may be grossly exaggerated, leading the designer to put more efforts to reduce power consumption at high-level. To alleviate this problem, we present a design automation solution to facilitate clock-gating at high-level. This will help a designer in controlling various granularities of clock-gating from the C description of the design.

We try to answer following questions in this chapter:

- Is there a way to enable RTL power aware trade offs from C specification?
- If answer to the question above is yes, can a designer/architect make such trade offs with existing automated HLS flow?
- If answer to the question above is yes, how to measure the efficacy of the approach?

We try to answer the problems/questions posed above by providing our perspective on the need for register clock-gating at high-level C description in Sect. 10.2. In Sect. 10.3, we discuss how to enable various granularities (such as function, scope, etc.) of clock-gating in C description, our algorithm, etc.

10.1.1   Main Contributions

The contributions of the presented methodology are following:

- Introducing a methodology implemented in a co-processor synthesis flow using C2R [41], for exploring power saving opportunities at the C-code level by

  - Exemplifying the complexities/issues related to assignment of clock-gating at a behavioral description level
  - Strategizing the selection of granularities of clockgating opportunities in a behavioral description written in ANSI-C
  - Demonstrating a specific priority algorithm to handle various conflicts for enabling clock-gating in functions, scopes, variables, etc.

- Comparison of the quality of the synthesized RTL (after such exploration) against the quality of results obtained by gate-level power optimization tools such as power compiler using a few industry strength benchmarks such as AES, GZIP, etc., design examples.

10.2   Why is Clock-Gating Needed at High-level?

Timing, power, or area can be controlled at architecture level by inserting directives in the C code or by applying intelligent scheduling and resource allocation schemes for synthesis. In case of power reduction, availability of tools to provide an estimation at the RTL with reasonable accuracy has opened up the opportunity to make macro/micro architectural power aware trade offs before RTL is finalized. Also clock-gating based power reduction sometimes leads to very high fidelity, it is very important to generate clock-gated RTL for these co-processors. One of the case studies presented in [115] shows that upto 70% of power reduction in design can be achieved with clock-gating alone.

Clock-gating is extensively supported at the RTL because designers have insight on how many registers are allocated to the datapath and controller of the design. If we compare it with respect to the C2R based co-processor synthesis, where only macro-architecture is fixed, enabling clock-gating is not as easy as it looks like. We introduce pragmas that can be inserted in behavioral code to enable clock-gating of registers during HLS. One should note that clock-gating can be enabled at different granularity from very coarse grain to fine grain such as register bank of 32-bit to a single bit register, etc. By macro architecture we mean the architecture where design contains minimal information about the implementation. Most of the times this design information contains various directives, processes to include parallelism and clock stages in the sequential C code. While the micro-architecture selection is done by the Cebatech C2R by applying design specific FSM to map the behavior of high-level specification.

In the high-level C description of the design, an architect can control the register power consumption by performing the gating for global/static variables. Similarly a case might arise to do the clock-gating of a register of a particular function (a hardware module at the RTL after synthesis) or even a scope of function. Such situations once handled at high-level will facilitate architect to control the register power of the design for various granularities. Design automation solutions, where HLS can be guided for register power reduction, will greatly help in providing early power-aware realistic design trade-offs.

Let us consider an example of pseudo C code as shown in Listing 10.1. In this example code a, b, c, and d are pointers to integers and foo is a function which is called by reference at two different places in the main function (line number 4 and 6). Now, Designer's/Macro-modeler's intent is to parallely execute the two foo modules in the generated hardware. After doing quick power estimation, designer comes to know that he just wants to clock-gate first foo block but not the second foo. There is no distinction between the two foo calls at the C description level. However, at the RTL two different foo modules can be instantiated as foo1 and foo2.

Listing 10.1 C source code to enable clock-gating

```
int  main  (  )
{
   int *a, *b, *c, *d;
   foo ( a ,  b )  ;
   . . . . . . . .
   foo ( c , d )  ;
   . . . . . . .
}

void foo ( int *x , int *y )
{
   int yyy ;
   int xxx ;
   . . . . . . . . .
}
```

Any C-based synthesis tool would require such an indication before generating RTL. On the other hand, to clock-gate various registers of foo1 and foo2 during logic synthesis designer just needs to provide information of these registers based on design hierarchy. In case of C description, we propose to use clock-gating(ON/OFF) directives, which will help the macro-modeler to control the granularities of various clock-gating decisions. Listing 10.2 shows the change in the code after applying the clock-gating directives. Details of different granularities and issues involved are available in Sect. 10.3.1.

Listing 10.2 C source code after inserting clock-gating directives

```
int  main  (  )
{
   int *a, *b, *c, *d;
   clock_gating (ON) foo (a , b)  ;
   . . . . . . .
   clock_gating (OFF) foo (c , d)  ;
   . . . . . . .
}

void foo ( int *x , int *y )
{
   int yyy ;
   clock_gating (ON) int xxx ;
   . . . . . . .
}
```

10.3 How to Enable Clock-Gating at High-Level?

It is very important that the high-level designer can exploit clock-gating benefits from the source code of the design. Techniques for power estimation presented in earlier chapters can help in finding out the hot spots of the design at high-level. These techniques can provide 6–20 times speedup as compared to the RTL power estimation techniques with acceptable accuracy. Once at the high-level, hot spots of the design are known, designer would like to gauge the impact of clock-gating on the power consumption of the design. One should note that such an analysis is performed across various components, and granularity may vary from a register-bank to a whole module. A high-level designer may just want to enable it at C description level because various scope/functions in the design are well known in the C description. Keeping this view in mind, below we present a few scenarios/situations that can be exploited from the C description to avail clock-gating at different granularities.

10.3.1 Application of Clock-Gating for Various Granularities at the Source-Code Level

For the C specification, designer can explore the granularities of clock-gating in many ways. Such an exploration is important because it helps the designer to take various decisions to reduce power consumption of a specific part of the design. At the C specification level, the designer has a visibility on various functions, scopes, variables, global variables, etc. Once designer knows which part of the design needs to be clock-gated, he can utilize directive such as clock_gating(ON) or clock_gating(OFF). Such directive used as macro in C environment won't affect functionality of high-level design in C simulation environment such as gcc [41]. It will direct the HLS to generate clock-gated RTL. Instead of using the directives, a script can also be used to direct HLS to clock-gate particular part of the design. Most of the times designer may want to clock-gate at coarse grain level such as scope, function (as discussed in Sect. 10.2). There exist some cases for fine grain decisions for which a careful analysis is required. We capture a few of these cases in this section and suggest a priority algorithm to take care of all the possible granularities for clock-gating. We discuss in detail how different directives can be applied for various granularities in the restructured C specification.

- *Function level*: As discussed earlier in the Sect. 10.2 (Listing 10.2) different calls of functions can be clock-gated ON or OFF using the clock-gating directives to get the function level control to clock-gate the design.
- *Scope level*: In a high-level description in C, another possibility to clock-gate the design comes at scope level. Sometimes, at scope level, designer wants to create hardware for the part of design (corresponding to the scope) and at the same time

he wants to clock-gate every register variable in that scope. Listing 10.2 shows designer can write a directive such as clock_gating(OFF) for not clock-gating a scope of function foo.

- *Variable level*: This is the simplest case in which user wants to enable clock-gating for particular variable. At compile time, if compiler sees a particular variable needs to be clock-gated because a clock_gating(OFF) directive is used so it will not clock-gate that variable, as shown in Listing 10.3.
- *Register bank*: Clock-gating every bit is sometimes very expensive because of its impact on area and timing. By clock-gating bank of registers such as 4 bits or 8 bits at a time, a lot of area can be saved because only one gating logic (a latch with and gate) can replace 4 and 8 multiplexers respectively.

10.3.2   Priority for Clock-Gating Decisions

As can be seen in the descriptions above, we enable clock gating before RTL is generated. In our approach, we provide the user with a flexibility so that he can turn clock gating ON or OFF at different granularities (e.g., at function level, scope level, variable level, or based on register bank size). Such provision without any priority will lead to conflicts. For example consider the example shown in Listing 10.3. In this example first call to foo contains the clock_gating(ON) directive while the other call contains clock_gating(OFF) directive. Even inside the function foo description clock_gating(ON) is used for yyy, while for xxx clock_gating(OFF) is used. Synthesis tool will see this as a conflict because for first call of foo, clock_gating(ON) directive is used, which suggests that all registers of foo (at line 12) will be clock-gated. On the other hand for xxx clock_gating(OFF) directive is used. In such a scenario high-level synthesis should know which directive should take precedence. We prioritize the clock-gating decisions based on the flow-chart shown in Fig. 10.1.

Listing 10.3 Example showing Conflicting Clock-gating directives

```
int main ( )
{
  int *a, *b, *c, *d;
clock_gating (ON) foo (a , b) ;
  . . . . . . .
clock_gating (OFF) foo (c , d) ;
  . . . . . . .
}
void foo ( int *x , int *y )
{
clock_gating (ON) int yyy ;
clock_gating (OFF) int xxx ;
  . . . . . . .
{
clock_gating (OFF) ;
  . . . . . . .
}
}
```

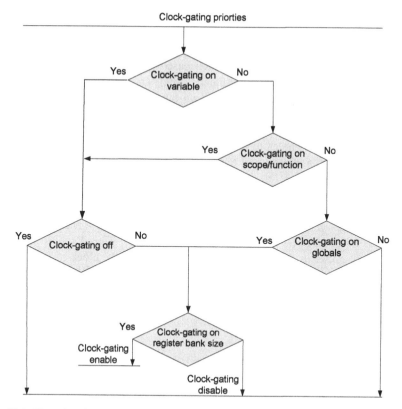

Fig. 10.1 Flow chart for gating clocks of registers at high-level

As shown in the flow chart in Fig. 10.1, if clock-gating is enabled/disabled for a particular variable then it will get the highest priority; then we check if the clock-gating is enabled/disabled at scope/function level and finally at global variable level. Finally,we check if the user sets a limit for register bank size, and if so, only those register-banks having size below that limit, will be enabled/disabled for clock-gating. In the example shown in Listing 10.3, for first call of foo, xxx will not be clock-gated while yyy will be clock-gated. Similarly for second call of foo, yyy will be clock-gated while xxx will not be clock-gated. One should note that high-level synthesis tool can be targeted with other priority but for our analysis we have kept the priority according to the flow-chart.

10.3.3   Algorithm for Enabling Clock-gating in C Based Co-processor Synthesis Framework

Our methodology takes advantage of the internal representation of C2R HLS tool. C2R compiler internally creates a Control Data Flow Graph (CDFG), which is passed through scheduling and resource binding after several optimizations. At the last pass when the registers in the design are fixed, we apply the algorithm for clock-gating. This enhanced scheduled CDFG is then used to generate the RTL Verilog containing gated clocks. We use the latch based clock-gating discussed in [58]. Our algorithm first finds out total number of registers in the design. Then it tries to find out clock-gating candidates based on the flow-chart or priority algorithm discussed in Fig. 10.1. Once registers/register-banks are finalized for clock-gating, our algorithm tries to find out the common enable conditions that are used to clock-gate the registers. This helps in further reducing the area (as discussed in [58]). Once the enabling conditions and registers are finalized, registers are changed to clock-gated registers.

To describe the algorithm we need to first fix a few notations as following: Let R be the set of registers in a design D; $NCG(r)$ be a predicate which is decided based on the flowchart in Fig. 10.1, and it evaluates to true if r needs to be clock gated, else false. Also, let $R_{cg} = \{r \mid r \in R \land NCG(r)\}$. Thus $R_{cg} \subseteq R$. Let C be the set of all clocks in D and $Clk(r)$ return the name of the clock of a register r. Let $Clkgated(r)$ return the name of the clock of r after clock-gating. For a register bank E having the same enable signal, $Clkgated(E)$ returns the clock of the entire bank after clock gating. Let En_r represent the enable signal for register r. Also, let $EN = \{En_r \mid r \in R_{cg}\}$.

Algorithm *clock-gating of registers*
Input : Scheduled CDFG in the last phase of HLS
Output : synthesizable RTL Verilog
Initially : $R_{cg} = \emptyset$

$\forall\, r \in R$ **do**
    if $NCG(r)$ then $R_{cg} = R_{cg} \cup \{r\}$.
    Find the enable signal En_r for r ;

Partition $R_{cg} = \{R_i^{Clk}\}_{i \in C}$
such that $R_i^{Clk} = \{r \mid r \in R_{cg} \land i \in C \land Clk(r) = i\}$.
R_i^{Clk} is the set of all registers having the same clock i.
$\forall\, R_i^{Clk} \subseteq R_{cg}$ **do**
    partition $R_i^{Clk} = \{E_j\}_{j \in EN}$
    such that $E_j \subseteq R_i^{Clk}$ and E_j is a register bank with all registers having the same
    enable signal En;
    $\forall\, E_j$ **do**
        Compute $Clkgated(E_j)$
        Replace clock (Clk) of all the registers in E_j with the $Clkgated(E_j)$.

Generate the hardware in Verilog (containing the gated clock for the intended registers).

10.4 Results

We present the results of our approach on a number of designs as shown in Table 10.1. We performed our experiments using our enhancement of a C synthesis compiler [41], PowerTheater [129] for RTL power estimation, and TSMC 180-nm libraries used in [144] for logic synthesis. Column 1 in the Table 11.1 shows the names of the designs experimented on, column 2 shows the power reduction achieved by applying clock-gating on every register using our approach, columns 3, 4, 5 represent the clock-gating applied on register banks of sizes greater than or equal to 4, 8, 16 bits, respectively. Table 11.1 shows that clock-gating on every register bit may not give the highest power savings. This is because our clock-gating introduces a *latch* and an *and* gate per enable signal, which also consume power. A trade-off point may be achieved by clock-gating register banks instead of individual registers. For example, in the caesar cipher case, almost 37% power reduction with respect to a non-clock-gated synthesis by individual register clock gating was achieved, but the savings further improved by clock-gating register bank of size 4 or more instead. NA in the Table 11.1 represents *not applicable*, because there was no register bank with size equal to or more than 16 bits.

We also compared the area and timing results of our approach against an automated RTL clock-gating technique using power compiler [120] shown in Table 10.2. In Table 10.2, we can see four cases where our approach found better power reduction opportunities, while impact on area and timing as compared to the lower level techniques is minor.

Table 10.1 Percentage power reduction as compare to the original design without clock-gating

Benchmark	% Power reduction using clock-gating			
	Every bit	4 bit bank	8 bit bank	16 bit bank
Fibonacci	16.32	16.41	16.32	NA
Caesar	37.78	38.22	38.22	NA
FIR	36.1	36.2	36.2	NA
AES	30.74	30.74	29.83	26.42
GZIP	16.73	16.39	14.96	12.29

Table 10.2 Comparison of our approach with automated RTL clock-gating

Benchmark	% Comparison of our results against RTL clock gating		
	Area	Timing	Power
Fibonacci	89.7	100	94.89
Caesar	112.10	100	91.03
FIR	100.2	100	100.2
AES	95.77	100	74.58
GZIP	100.36	100	100.36

For example, for AES design, about 3% area saving was obtained while there was no impact on timing as compare to the results generated by power compiler. Our approach shows almost 25% extra power savings as compared to power compiler. This shows that our algorithm exploits the visibility of the registers and their control inputs especially for bank or registers. The purpose for this table is to provide a view to the reader that using our approach a very realistic design exploration is possible at the high-level in shorter time, without compromising quality, and some times with more power savings. In this table, we normalized the values from power compiler as 100.

10.4.1   Clock-Gating Exploration at High-Level

In the Sect. 10.4, we discussed how a designer can filter the clock-gating register candidates and compared the results of our approach with respect to RTL clock-gating techniques. Here, we present a few example scenarios that can be easily explored at high-level. To explore the same scenario at the RTL, designer would require a lot more information from HLS to correctly clock-gate registers. It is expected that the person working at C source level would have more detailed knowledge about the architecture, and micro-architecture is produced by the synthesis tool. Hence, relying on the RTL clock-gating techniques would increase the design time.

Let us consider Fibonacci example. As we know, in Fibonacci series, the number n is provided and corresponding series is emitted by the design/software. Now, depending upon the architecture/constraints on the design for a number n, design may take several clocks to produce the series. We noticed that n remains constant till design creates complete sequence. Generally, value of n changes based on the input signal which suggest that data input is valid on the port (using an interface function). This suggests clock-gating this variable and other variables/scopes that are using this variable may lead to good power savings. Similarly, in the case of AES design, it is a symmetric cipher and the *key* to encrypt/decrypt data remains same for several data iterations. Hence, it make sense to just clock-gate variables saving value of key and the variables using value of key, as there might not be a lot of toggling happening on these registers. Now, applying such changes for clock-gating is very easy at source code level using our technique.

As discussed in Sect. 10.3, designer would require to put a directive to guide HLS to clock-gate the registers we are interested in. We observed almost 10% power savings for fibonacii example while in case of AES we could see little less than 5% power savings. In AES design, there are a few functions which can be separately clock-gated to see which function has the maximal impact on power consumption of the design. Such optimizations if applied at RTL would require a lot of effort especially in an HLS based design flow. In these cases effect on timing was neglige and we observed little saving in the area as well because of replacement of mux for a register bank with clock-gating logic.

10.5 Summary

In this chapter, we answered a few questions discussed earlier on enabling clock-gating in ANSI C description using clock-gating directives. This will guide HLS to generate power aware hardware. We first show how different directives can enable clock-gating at high-level starting from a variable, fixed bit-width, scope to a function. Secondly, we provided an algorithm on how such directives would be taken care by the HLS tool. Finally, we provide example scenarios that an architect can easily see in the high-level description. Enabling such reductions at the RTL in HLS based design flow would require back and forth interaction between HLS and lower level tool, making the power aware exploration task very tedious and at the lower-level of abstraction.

We implemented the strategy into a commercial co-processor synthesis tool and applied it to synthesize a number of co-processors. We show that a significant power reduction is possible, sometimes beyond what a lower level automatic clock-gating tool would provide, without compromising area and timing. We also present various ways of controlling the clock-gating granularity. Precise control on various variables in ANSI C description helps in getting rid of redundant clocking of the registers of the design for different granularities.

Chapter 11
Model-Checking to Exploit Sequential Clock-Gating

Dynamic power reduction techniques such as sequential clock-gating aim at eliminating inconsequential computation and clock-toggles of the registers. Usually, sequential clock-gating opportunities are discovered manually based on certain characteristics of a design (e.g. pipelining). Since manual addition of sequential gating circuitry might change the functionality of the design, sequential equivalence checking is needed after such changes. Tools for sequential equivalence checking are expensive, and based on recent technologies. Therefore, it is desirable to automate the discovery of sequential clock-gating opportunities using already existing and proven technologies such as model checking and thereby a priori proving that the changes will not affect the required functionality. Model Checking Based Sequential Clock Gating (MCBCG) method formally proves particular sequential dependencies of registers on other registers and logic, thus sequentially gating such registers will not require further validation. An automation scheme for MCBCG methodology is also proposed in this chapter. Our experiments show up to 30% more savings than the traditional (combinational) clock-gating based power reduction techniques. We further experimented the approach to find out the savings opportunities exist in HLS tools. This further suggests that at high-level finding such opportunities might be very benefical for HLS based design flow.

11.1 Introduction

Register-power is generally one of the biggest contributors [58] to the total dynamic power consumption. To obtain maximal power reduction of the design, it is advisable to reduce the register power of the design. Clock-gating [58] is one of the techniques which has been extensively used to reduce the dynamic power consumption. The main idea behind clock-gating is to reduce the unnecessary toggles of registers when the register update is not required or its value is irrelevant to the computation. In the absence of clock-gating, at every clock cycle, each

S. Ahuja et al., *Low Power Design with High-Level Power Estimation and Power-Aware Synthesis*, DOI 10.1007/978-1-4614-0872-7_11, © Springer Science+Business Media, LLC 2012

register gets updated even when it remains unchanged in value. One can utilize the enable/control signal for a register to stop the clock from making the register toggle. The computation of the enable/control signal to indicate that the clock needs to be stopped can be done in two ways. When the information is temporally localized and only based on the current cycle, it is known as *combinational clock gating* described in the Chap. 10. However, more benefits can be obtained if it is known that the register value is not going to be used in the future cycles, or if certain events from some previous cycles have indicated that register update is unnecessary. Utilizing such information to gate a register's clock is commonly known as *sequential clock-gating* [105].

Keeping such information across clock cycles requires relevant circuitry to propagate the information temporally into future cycles. Also, knowing whether register's value update is irrelevant to future computations depends on the dynamic relationship between the register and other events during the execution. If this information can be inferred from the structure of the design, then designers can manually insert the relevant gating circuitry. In pipelined designs, these kind of inferences are often not that difficult [97]. It is hard, when the structure of the design does not provide an obvious opportunity, or when the designer may suspect that such relationships across clock cycles exist, but cannot be sure due to the complexity of the design. Tracing the fanin or fanout cones of registers can be often helpful, but requires automated analysis [15].

In any case, such optimizations that can affect the functionality of the design over the span of time/clocks require considerable effort on verification, usually best done by checking sequential equivalence. In traditional methodologies, first these optimization techniques are applied manually and then equivalence checking is performed. However, sequential equivalence checking technologies are new, and hence using a more proven technology would be more desirable. Keeping this in mind, we propose a novel methodology called as *Model Checking Based Sequential Clock-Gating (MCBCG)*. MCBCG utilizes model-checking to apriori prove that sequential clock gating opportunity exists for a register r, based on other registers, whose activities in past cycles would always indicate irrelevance of updation of this register in a future cycle. Since we are using model checking, we can try to prove this for any arbitrary set of registers if we wish, and hence we do not depend on the special structures such as pipelines. Of course, designer should apply this in a more informed manner than arbitrarily. Note that most of the current approaches we found in the literature, implement sequential clock-gating based on the architecture of the design such as pipelined designs. In contrast, our approach and the one presented in [15] can be applied to any RTL model of the design. Secondly, since we already prove that sequentially clock gating the registers found using this approach would not affect the computation, we do not need expensive sequential equivalence checking afterwards.

Main contributions of this work for the presented methodology are as follows:

- A novel methodology to investigate the sequential clock-gating opportunities in hardware designs.

- Demonstration of some temporal properties that can be used for model checking to infer these opportunities.
- Experimental results showing opportunities to save substantial power over combinational clock gating based power reduction.
- Extending the HLS framework to generate sequentially clock-gated RTL.

11.2 Our Approach and Sample Properties

11.2.1 An Illustrative Example

We briefly discuss here a few possible scenarios that will help in understanding the rationale for our approach. Let us assume two register a, b and b is dependent on a. In Snippet 2, $in1$, $in2$ and en can be considered as input. Snippet 3 shows how the values of the registers change. We can see from this example that if value of a is not utilized elsewhere, then a is not required to do the unnecessary computations.

Snippet 2   Representation of two registers p, r

```
.....
next(a) = in1 + in2;
.....
if(en)
    next(b) = a + 5;
```

Snippet 3   Clock by clock view of the changes in value of various registers

```
Initially a = 0; b = 0;
cycle 1 in1 = 5, in2 = 5; en = 0; a = 0, b = 0
// results are updated a cycle later
cycle 2 in1 = 7, in2 = 6; en = 1; a = 10, b = 0
// b gets updated because en was true
cycle 3 in1 = 8, in2 = 7; en = 0; a = 13, b = 15
// en was false, b won't change
cycle 4 in2 = 7, in2 = 2; en = 0; a = 15, b = 15
```

Similarly, let us consider the example where a can be clock-gated while b cannot be but they have a relationship as shown earlier. Snippets 4 and 5 show register b gets unnecessary clock-toggles since there is no value change but still clock is supplied. In this case, after applying sequential clock-gating, functionally design would not exhibit any change but dynamic power still can be reduced. While in the earlier example, if value of register a is used in other parts of the design then it is necessary to check that value change in the register a should cause a value change on some other register after a clock-cycle. If it is the case we can't do a sequential clock-gating. While for the second case we can obviously apply sequential clock-gating because it will not affect the design functionality.

Snippet 4   Example of operations on registers a, b

```
.....
if(en)
    next(a) = in1 + in2;
.....
next(b) = a + 5;
```

Snippet 5   Clock by clock view of the changes in value of various registers

```
Initially a = 0; b = 0;
cycle 1 in1 = 5, in2 = 5; en = 0; a = 0, b = 0
// results are updated a cycle later
cycle 2 in1 = 7, in2 = 6; en = 1; a = 0, b =5
// en was false, a won't change
cycle 3 in1 = 8, in2 = 7; en = 0; a = 13, b = 5
// a gets updated because en was true
cycle 4 in2 = 7, in2 = 2; en = 0; a = 13, b = 18
// b changed a cycle later, a did not change because
// en was false
```

11.2.2   Our Approach

As discussed in Sect. 3.8.2, sequential clock-gating can be applied if two registers are dependent on each other in certain ways. One way to find these dependencies is to go through the structure of the design and find out such relationships (explained later in this chapter). This requires traversal of the netlist. However, in some cases, for example, in the case of Snippet 2, it cannot be inferred through such traversal. Fortunately, by utilizing a model-checker we can find such intra-register relationships across clock boundaries, both in the case of structurally inferable dependences, and data dependent relations. Based on the discussion in Sects. 3.8.2 and 11.2.1, we can map this relationship to a property of the design. Such a property is basically run as a query to the database of the design, asking if a relationship between registers exist or not. Once this property is verified by the model checker and the answer to the query is "yes." Then we know how to exploit this opportunity to utilize sequential clock-gating in the design.

Let us assume we have the implementation model and a formal model (for model checking) for a design under consideration. Some design groups in industry/academia utilize tools that can convert the RTL description to the formal models for model-checking purpose. The formal model must capture all the registers and register transfer logics from the implementation model. Now we will see some sample properties that one can write on the formal model. If these properties are verified, then we can use the implied relationship for clock-gating the registers, where redundant activities are happening. Let us consider two arbitrary registers r, p in the design. Suppose the value of p changes at an arbitrary clock cycle t, and the value of r changes at clock cycle $t + 1$. We want to verify if r changes at cycle t, then p must have changed at cycle $t - 1$. In the formal model, we add three extra state variables for register r and p. Let us name them as p_pre2, p_pre and r_pre with the same types as p and r. Initialize p_pre2, p_pre with the initial value of p and r_pre with the initial value of r. Then make sure to add the transitions for p_pre2, p_pre and r_pre. We will show the transitions using the syntax of Cadence SMV [38] for encoding value changes of registers. In Cadence SMV, the transition of register s is written as shown in Snippet 6.

Snippet 6   Registered value representation in Cadence SMV

```
next(s) = expression;
```

The expression may have if-then-else to encode the multiplexer logic. When the formal model has registers r and p, then it will look as shown in Snippet 7.

Snippet 7   Representation of two registers p, r

```
next(p) = expression1;
next(r) = expression2;
```

Change the model shown in Snippet 7 to the one shown in the Snippet 8.

Snippet 8   Changes in the model to verify the properties

```
next(p_pre2) = p_pre;
next(p_pre) = p;
next(p) = expression1;
next(r_pre) = r;
next(r) = expression2;
```

Therefore, at cycle $t + 1$ p_pre, r_pre contains the value of p and r from cycle t. So if $(p\_pre \sim= p)$ is true that means at cycle $t + 1$ value of p changed. Similarly, if $(p\_pre2 \sim= p\_pre)$ at cycle $t + 1$, that means at cycle t p changed from its previous value at cycle $t - 1$. Let us consider the following two properties:

1. At cycle t, if r changes, p must have changed in $t - 1$. Verify the property shown below. If this property is true, we can see if p can be clock-gated then r is a potential candidate for clock-gating in the next cycle.

$$G((r\_pre \sim= r)- > (p\_pre \sim= p\_pre2));$$

2. At cycle $t - 1$, if p changes, then r must change at cycle t. Verify the property shown below. If this property is true then we have to check if the value of p is not used by some other register and r can be clock-gated then p is a potential candidate of sequential clock-gating.

$$G((p\_pre \sim= p\_pre2)- > (r\_pre \sim= r));$$

By checking these and other similar properties on the formal model, we can be sure that de-asserting enable of p at any cycle t, would imply that r will not change its value at $t+1$. Hence, we can apply sequential clock-gating for the related registers r and p. G is the linear-time temporal logic (LTL) operator used in the example listings to represent always. We can write properties till n clock cycles, where n is the number of clock stages we are interested in. We call this as window size. Practically, this number won't be too big, because every time n increases we introduce extra registers and clock gating logic to reduce the power consumption, which may affect area largely.

11.3   Our Proposed Methodology

As shown in Fig. 11.1, MCBCG methodology starts with the formal model of the design. Once such a model is available, we can start writing the properties of the design as discussed in the Sect. 11.2.2. Then pass the model along with the properties (queries) to be verified to a model checker. We have used Cadence SMV for our analysis but other model checker such as VIS [154], ABC [2] can also be utilized for the similar analysis. VIS, ABC can directly read in the RTL model and facilitate user to write properties on the implementation model. We use SMV because of the familiarity with the tool.

Once the properties are verified by the model checker then based on the relationship between the registers apply the power reduction specific changes. These changes requires some changes in the RTL verilog and very simple to implement at RTL. In this paper, we applied latch based clock-gating scheme. The advantage of such a scheme is that combinational glitches can be avoided in the clock port of the register. If property is false then apply the similar property to other set of registers. Once RTL model is ready with the applicable changes for sequential clock-gating, utilize this model for power estimation. One should note that we apply clock-gating using power compiler in both the models but the difference is the enabling condition for the registers which can be sequentially clock-gated in our approach. This whole process can be automated by first finding out the structural relationship between the register pair and then utilizing our technique to find out if register pair are good candidates for clock-gating based scheme. Finally once the power optimized RTL is ready, we validate whether our approach provide any further reduction or not.

These reductions may show very good results for one set of stimulus and in other cases it may not. It is very important to choose right stimulus for the analysis. To measure the differences with respect to the original model, power consumption for both the models is estimated using same technology libraries and the same test vectors. Finally the difference in area, timing and power can be measured after completing the design synthesis, etc.

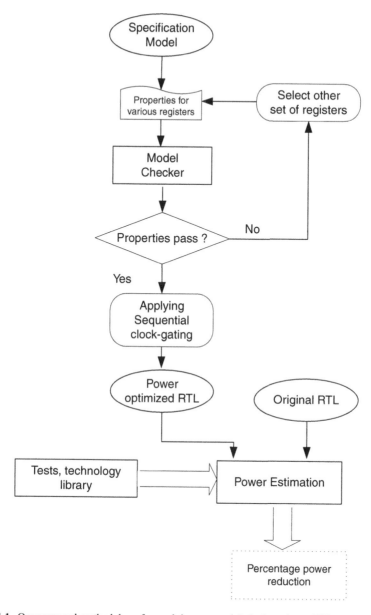

Fig. 11.1 Our proposed methodology for applying sequential clock-gating at RTL

11.3.1 Changes Applied on the RTL Model

We propose minimal changes for the RTL modeling stage. The code changes are shown using Verilog hardware description language (hdl), similar changes would be required with VHDL or any other hdl. After including these changes tools such as power compiler can directly be utilized. Such a facility will help the designer to stay in the current design flow. Listings 11.1 and 11.2 show the changes that can be applied to the code. In Listing 11.1, we show how two different registers q, $q1$ may look like in Verilog. After doing the analysis using MCBCG methodology, we know that these two registers are the candidates for sequential clock-gating. We utilize the enabling logic of the two registers. One should note that a change in value of $en1$ a cycle before is captured first in $en1\_tmp$ register. Finally we utilize the combinational logic of the two register's i.e. $en1\_tmp$, $en2$ and store it in a wire (using assign statement in Verilog) and change the $en2$ with $en2\_tmp$. All code changes are illustrated in Listing 11.2. We can also include the clock-gating specific changes directly in the RTL model but for our experiments we utilized power compiler to apply combinational clock-gating.

Listing 11.1 Verilog code segment before applying MCBCG technique

```
// register q with enable en1
always@(posedge clk or negedge rst)
begin
// Reset logic skipped
  if (en1)
    q <= r;
......

end

// register q1 with enable en2
always@(posedge clk or negedge rst)
begin
// Reset logic skipped
  if (en2)
    q1 <= r1;
......

end
```

Listing 11.2 Verilog code changes after applying MCBCG technique

```
// register q with enable en1
always@(posedge clk or negedge rst)
begin
// Reset logic skipped
  if (en1)
    q <= r;
......

end

always@(posedge clk or negedge rst)
begin
// Reset logic skipped
// this will delay the en1 by a clock
// and latch it in a register
    en1_tmp <= en;
......

end

// New enable dependent on the previous
// enable condition
assign en2_tmp = en1_tmp and en2;
```

```
// register q1 with enable en2
always@(posedge clk or negedge rst)
begin
// Reset logic skipped
  if (en2_tmp)
    q1 <= r1 ;
......

end
```

11.3.2 Automation of the Flow

Algorithm 9 presents a scheme to automate the methodology. To describe the algorithm we need to first fix a few notations as follows: Let R be the set of registers in a design D; W be the window size (different stages of the registers to find the clock-gating opportunities), T be the set of register bank of the same type, and P be the set of properties for every register pair.

Algorithm 9 Automation of *sequential clock-gating of registers*

Procedure SeqCG{
$\forall\, w \in W$ **do**
   execute Procedure WindowCG

}
Procedure WindowCG{
Find out all register bank of same type (bitsize) t
$\forall\, t \in T$ **do**
   collect all register pairs
   Find out the relationship between the registers using properties discussed in Sect. 11.2.2
   $\forall\, p \in P$ **do**
     if property holds
     For both the cases discussed in Sect. 11.2.2 find out candidate register pairs for sequential clock-gating
     collect all the unique register pairs in a list

   Apply the sequential clock-gating optimization and change the verilog as discussed in Sect. 11.3.1 for each register pair of the list

}

In the above procedure based on the window size, every property is verified for register bank of the same type. Doing such a task requires a knowledge of the register bank and would require addition of extra register stages in the formal model. This task can be incrementally performed for all the registers in the design. We have implemented the procedure using c script. Once property is verified on a register bank we save this information in a separate list, from there we finally post process the original RTL based on the steps discussed in Sect. 11.3.1.

11.4 Results and Conclusions

Our investigation in this work show promising results. We tried the approach on two sample designs FIR filter and three-stage pipelined packet processing design. We could achieve more power savings than combinational clock-gating using power compiler. Table 11.1 shows the comparison of our approach against power compiler. Our approach shows that almost 30% more power savings can be obtained. We used VCS tool for RTL simulation [149], power compiler [120] for clock-gating and power estimation, and 180nm libraries from [144].

11.4.1 Extending the Framework for HLS

This chapter presents application of model checking to find out a relationship between registers. However, model-checker may not always be available for HLS based framework. In this section, we briefly present the steps used for structural traversal for sequential clock-gating opportunities. As discussed earlier, sequential clock-gating technique relies on the fact that enable condition of one register a cycle or two cycles before can be utilized in earlier/later stages of the design. These steps are captured in Snippet 10 to perform structural traversal of the intermediate Control/Data Dependency Graph (CDFG).

11.4.1.1 Results

The steps presented in the earlier section show the structural traversal of registers on the CDFG of the design. As discussed earlier in Chap. 2, model-checking based approach can help in finding some of the behavioral relationships which structural approach may not find. In our experiments, we first verified the generated designs for all the benchmarks. We used simulation based approach where correct functionality of the design was verified with and without sequential clock-gating for the same

Table 11.1 Percentage power reduction summary on the tested designs

Design	Total power consumption	Power consumption after c.g.	Power consumption after seq. c.g.
FIR	1,106 uW	522 uW	426 uW
Three-stage pipe	301 uW	239 uW	147 uW

Snippet 10 Steps for finding sequential clock-gating opportunities at high-level

---

Step1: Obtain the dependency graph of the design and identify the registers of size atleast 4 bits and 8 bits (or bank size specified by the user). In sequential clock-gating an extra register is introduced for sequn, so this scheme needs to be implemented for a register-bank.

Step2: Save every register and its dependency on other registers and mark the dependency as levels while traversing the control flow.

Step3: foreach (level)

  Apply Step 4;

  Step 4:

    a-) From the dependency of every register from step2, find out the relationship of every register for the future stages

      b-) If the register in the future stages are dependent on more than one current state registers than use a combinational logic of all the enable conditions and delay it to clock-gate future register stage

      c-) else delay the gating logic by a clock using a register

end foreach

Step 5: Generate the Verilog RTL of the new design

Table 11.2 Percentage power reduction summary at high-level

Design	Total power consumption	Power consumption after c.g.	Power consumption after seq. c.g.
AES	123 mW	90.6 mW	89.2 mW
Three-stage pipe	1.11 mW	1.16 mW	0.752 mW
FIR	2.9 mW	1.85 mW	1.79 mW

test-vectors. We observed that every benchmark has not shown the opportunity to sequentially clock-gate because of the nature of the design. In our experiments, we found that designs having a pipelining nature or direct dependency between two registers have shown some power savings and could be optimized further. Table 11.2 captures the results obtained.

Chapter 12
System Level Simulation Guided Approach for Clock-Gating

Clock-gating is a well known technique to reduce dynamic power consumption of a hardware design. In any clock-gating based power reduction flow, automatic selection of appropriate registers and/or register banks is extremely time-consuming because power analysis is performed at the RTL or lower level. In a high-level synthesis (HLS) based design flow, to achieve faster design closure, one must be able to decide the appropriate set of registers to clock gate even before generating RTL. System-level simulations are known to provide faster simulation, yet there is no solution, which utilizes system-level simulation to provide guidance to HLS to create clock-gated RTL. Since predicting power reduction at higher levels of abstraction is difficult due to the dependence of power on physical details, an accurate and efficient *relative* power reduction model is required. In this chapter, we propose a novel system-level design methodology, which utilizes a "relative power reduction model" that can help in predicting the impact of clock-gating on each register/bank quickly and accurately, by simulating the design at a cycle accurate transaction-level. As a result, our approach can automatically find the appropriate registers to clock-gate, guided by the system-level simulation.

12.1 Introduction

Currently, most of the HLS based design flows include power reduction pass through RTL or gate-level tools such as Power compiler [120], PowerTheater [129], etc. However, these tools rely on designers' insertion of clock-gating directives for the registers at the RTL. In other words, it is the designer's responsibility to select if all registers should be clock gated, or registers belonging to specific modules or specific parts of the design should be clock gated. If a register is updated very frequently with new values, inserting clock gating logic causes unnecessary overhead with adverse effect on power consumption. The other impact is on the timing closure.

S. Ahuja et al., *Low Power Design with High-Level Power Estimation and Power-Aware Synthesis*, DOI 10.1007/978-1-4614-0872-7_12,
© Springer Science+Business Media, LLC 2012

If unnecessary clock-gating is done, more timing issues will arise at the lower levels of abstraction. Clock-gating should be applied only if a significant amount (i.e. above a certain threshold) of power savings for every register can be observed [64].

A few aggressive techniques to perform clock-gating utilize observability don't care (ODC) to find signals whose value change is don't care, and hence can be gated [15, 63]. However, the power consumption by additional circuitry can overwhelm the savings if such a technique is applied for every register. Most recent work such as [15, 63] suggest the registers for which more power saving opportunities may exist but cannot determine from the simulation if clock-gating based reduction will provide savings for every register/register-bank. The savings are presented on an average for the whole design and at the netlist level. Furthermore, to be effective at high-level, this information should be available even before RTL is generated.

In this chapter, we show that system-level simulation can be utilized to generate clock-gated RTL using HLS. To achieve this, we propose a novel power reduction model, which measures the activity of a register value change at cycle-accurate transaction level and then guides the HLS to generate the clock-gated RTL. This approach helps in finding whether each register would save or cost power if an aggressive power reduction approach is used in the design flow. We have benchmarked the accuracy of our estimation of power savings on the industry strength co-processor design blocks, and found that the accuracy of our estimation is encouraging (approx. 7%) and speedup to perform power analysis is in 2–3 orders of magnitude. We also show that for some designs, upto 45% of the registers in the design may not require clock-gating.

12.2   Power Reduction Model

We present power reduction model for clock-gating in this section. As we know the idea behind clock-gating is to reduce the unnecessary clock toggles. We introduce two power models in this section and the reason comes from the availability of technology libraries in the early design flow. If early in the design flow, good technology libraries are not available and yet we want to measure the impact of power reduction on the generated RTL. For such a scenario, extraction of effective clock updates from simulation of clock-gated design is required. Section 12.2.1 provides detailed analysis of this approach. A lot of times, during high-level synthesis, technology library information is available. In that case we can enhance the power model with such information and get a better visibility on the reduction that can be achieved. In that regard, we require information about integrated clock-gating cell and multiplexers. Section 12.2.2 provides details of technology dependent power model.

12.2.1   Activity Based Power Reduction Model

Our relative power model depends on "savings duty cycle" estimation. In our HLS flow, cycle accurate transaction level (CATL) and RTL models are generated from the same control flow graph (CFG); hence, the correct information on enables and registers are available while generating the CATL model. Let us assume that the enabling condition for a register update is denoted by en. To extract the useful update of a register, we insert a counter for each register of the design in the CATL model. Snippet 11 shows a very simple code after including a counter for every "enable" of a register in the design. This counter helps in capturing all the effective updates happening to a register/register-bank (in this case, register q). This model can be simulated in a SystemC simulation environment.

Snippet 11   Counter inserted in the generated SystemC code

```
if(en){
    q = some computation;
    // counter for q will be inserted here
    counter_en = counter_en + 1;
}
```

Let us divide the total dynamic energy (E_{total}) for a register into two parts E_{useful} and E_{waste} (represented in (12.1)). Let R be the set of registers in a design D; let N_r represent the inserted counter for "enable" of each multiplexer in front of a register r and N_{clk} a counter for the clock. E_{useful} can be represented as shown in (12.2). Similarly, E_{waste} can be represented as shown in (12.3). k is a constant of proportionality dependent on area, capacitance etc.

$$E_{total} = E_{useful} + E_{waste} \tag{12.1}$$

$$E_{useful} = k \times (N_r/N_{clk}) \times V^2 \times f_{clk} \times T \tag{12.2}$$

$$E_{waste} = k \times (1 - (N_r/N_{clk})) \times V^2 \times f_{clk} \times T \tag{12.3}$$

Thus, $(1 - (N_r/N_{clk}))$ is the "savings duty cycle" for the register r. This power model can help us in evaluating the impact of clock-gating on power savings. If the information on technology libraries and clock-gating cell is not available, designer can provide a threshold to select the clock-gating candidates as shown in (12.4). This inequality checks if savings are above certain limit ($a_{threshold}$). If it is, then clock-gate the register. Since the number of registers before and after clock-gating are same, such a relative power model will help in finding out the power reduction opportunities even before applying the clock-gating changes in the RTL. $a_{threshold}$ is a fraction between 0 and 1.

$$(1 - (N_r/N_{clk}) - a_{threshold}) \geq 0 \tag{12.4}$$

12.2.2 Power Reduction Model with Technology Specific Information

Roughly, the power model presented in earlier section shows how much clock toggle savings exist in a design. In this section, we propose a more detailed power model of the design to aid the analysis. It is well known that when clock-gating is applied to a design then only modification in the design comes by removing a multiplexer from the feedback path of a register; also a clock with gating logic is supplied to the clock port. Our power model captures these changes. We utilize an integrated clock gating cell (icgc, generally the logic contains an *and* gate with a latch) to provide the gated clock. For the sake of clarity, we limit our discussion to single bit register bank but for a register bank of size N and icgc of fanout strength p, a few minor modifications would be required. As we know that power consumption is dependent on capacitance, which further depends upon the area; and the area information of these cells can be extracted from the technology library to improve the accuracy of the power model.

In our case, we used standard cell library from TSMC and obtained area information (details on our experimental setup are provided in the results section). Generally, all the standard cell technology libraries contain this information. One should note that we do not require information for all the cells rather a few cells, which are intended to be utilized for clock-gating purposes only. Equation (12.7) shows the coefficient, which is computed for various simulation vectors, if this coefficient is positive, tool will clock-gate the register, else it will not. The negative number shows that there would not be any power savings using clock-gating for a particular register. This way tool will automatically infer the information of registers to be clock-gated directly after the simulation.

Let us represent the energy of a register, mux, and icgc as E_r, E_{mux}, and E_{icgc} respectively. Frequency of updates for register and mux can be denoted as N_r and for icgc N_{clk}. Dynamic energy of a register before and after applying clock-gating can be denoted by E_{r1}, E_{r2} as shown in (12.5) and (12.6).

$$E_{r1} \propto A_{reg} \times N_{clk} + A_{mux} \times N_r \tag{12.5}$$

$$E_{r2} \propto A_{reg} \times N_r + A_{icgc} \times N_{clk} \tag{12.6}$$

Difference in the energy values (which represents energy savings/wastage before and after clock-gating) with respect to energy of a register before clock-gating can be represented as shown in (12.7):

$$p_c = \frac{E_{r1} - E_{r2}}{E_{r1}} = 1 - \left(\frac{N_r \left(1 - \frac{A_{mux}}{A_{reg}}\right) + N_{clk} \times \frac{A_{icgc}}{A_{reg}}}{N_{clk}} \right) \tag{12.7}$$

Equation (12.7) clearly shows p_c will not always be positive but it depends on a lot of factors. This equation also shows that introduction of an icgc may cause power loss because it is always fed with the 100% activity of the clock port. This is also the point in the clock-tree from where the clock is branched and duty cycle is reduced. On the other hand, there will be some savings because of removal of muxes. In case of register banks, the power savings would be higher because an icgc would be introduced for the bank but its drive strength would be computed based on the size of the register bank. System-level simulation can make the process of computing N_r, N_{clk} very fast. Values of A_{mux}, A_{reg}, A_{icgc} will change according to the standard cell library used in the flow. For N bit register bank and p bit fanout of an icgc the power model will look as shown in (12.8). After including the threshold provided by the designer, relative savings look as shown in (12.9).

$$1 - \left(\frac{\left(N_r + \left(\frac{A_{icgc}}{A_{reg}}\right)\right) \times N_{clk}(1/N + 1/p) - N_r\left(\frac{A_{mux}}{A_{reg}}\right)}{N_{clk}} \right) \geq 0 \qquad (12.8)$$

$$p_c - a_{threshold} \geq 0 \qquad (12.9)$$

12.3 Rationale for Our Approach

Let us consider that for a design (at RTL) R and for vector i simulation takes T_i^R time. Let us represent the clock-gated RTL as R_c and the time to simulate vector i as T_i^{Rc}. Let C represent different configurations such as fine grain clock-gating for every register, register-bank, or to a particular module (coarse grain), etc. To check the efficacy of power reduction technique on the design for n different simulation vectors, the power estimation time required T_E^c can be calculated as per the simple procedure shown in the Procedure PowerEstTime.

Procedure PowerEstTime() {

$\forall\, c \in C$ **do**
  $\forall\, i \in n$ **do**
    Simulate RTL and Estimate power for the design R, R^c

}

Equation (12.10) represents the total time to simulate RTL with and without clock-gating for each simulation vector as per the procedure *PowerEstTime* discussed above. T_i^R, $T_{E_i}^R$ represent the time taken in performing simulation and power analysis for the design R before applying clock-gating, and T_i^{Rc}, $T_{E_i}^{Rc}$ represents the same after applying clock-gating.

$$T_E = \sum_{c \in C} \sum_{i=1}^{n} T_{E_i}^c = \sum_{c \in C} \sum_{i=1}^{n} (T_i^R + T_{E_i}^R + T_i^{Rc} + T_{E_i}^{Rc}) \qquad (12.10)$$

Let us assume, there exists a system-level approach that can help in measuring the power savings using high-level power model (such as the one discussed in the last section) and analysis time remains close to system-level simulation and can be represented as T_i^s for vector i and total system-level simulation time can be represented as T^s. As we know, system-level simulations are faster than RTL, we can assume:

$$T_i^s \ll T_i^R \tag{12.11}$$

Experimentally, we found that cycle accurate transaction level (CATL) simulation is at least 10–15 times faster than corresponding RTL simulation. This can be attributed to quite a few reasons, for example, CATL simulation is applied to data-types such as char, integer as compared to bit-level data-types in RTL simulations. Also one can use functions, pointers, which makes the simulation faster than the RTL simulation. Relationship between system level and RTL simulation time can be represented as shown in (12.12).

$$\frac{\sum_i T_i^R}{\sum_i T_i^s} = \mu \tag{12.12}$$

As per the discussion above

$$\mu \gg 1; \tag{12.13}$$

The ratio of RTL and system-level power analysis time can be represented as

$$\frac{T_E}{T^s} = \frac{\left(\sum_i (T_i^R + T_{Ei}^R) + \sum_c \sum_i T_i^{R_{ic}} + T_{Ei}^{R_c}\right)}{\sum_i T_i^R / \mu} \tag{12.14}$$

$$\mu + \frac{\mu\left(\left(\sum_i T_{Ei}^R + \sum_{c \in C} \sum_i (T_i^{R_c} + T_{Ei}^{R_c})\right)\right)}{\sum_i T_i^R} \tag{12.15}$$

At the RTL, power estimation tool would need to extract activity from the simulation dump and map it to the design for getting power numbers. This is one of the most time consuming processes of a power estimation tool. Based on our experience, we find that all these factors in an RTL power analysis tool are almost 10–100 times slower than the system-level simulation [11]. This would make the speedup in the range of 2–3 orders of magnitude. We experimentally verify this claim in the results section. The rationale to use system-level approach can dramatically improve the design cycle time from the low power design perspective.

12.4   Our Methodology

A typical HLS based design flow involves generation of RTL, insertion of clock gating based on user inputs at the RT level, followed by logic synthesis, and then a pass through power estimation tools. These estimation tools require design information in verilog/vhdl, simulation dump in vcd/fsdb formats, and technology library information (.lib), etc. Power estimation is very time consuming mainly because RTL/gate-level simulation is slow and the generated vcd/fsdb is large and needs re-processing for power estimation purposes. After this time consuming single pass, if power closure is not achieved, another time consuming pass is made, and such passes are repeated until power closure is achieved. Figure 12.1 illustrates the iterative process in a standard HLS enabled design flow.

We discuss in detail some of the distinct features of the methodology proposed in this chapter. These features include integration of tool generated SystemC model with un-timed TLM model to achieve very high speed analysis. Second important

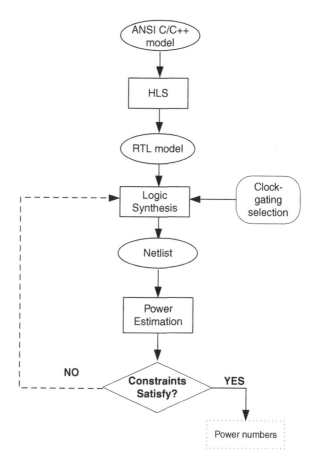

Fig. 12.1 Traditional clock-gating methodology for HLS based design flow

part is the support of clock-gating at ANSI C based description itself. Section 12.4.1 provides a brief overview of our interconnection framework. Chapter 10 provide details on how to facilitate clock-gating from the C description itself. The advantage of providing the control from C description comes by controlling granularities of clock-gating, which RTL clock-gating tools can not find by netlist traversal.

In today's design flow, designer applies clock-gating based on their experience and there is no way in which he/she can measure individual register's power savings in the design. One possible solution is to generate power report for each register's power consumption before and after applying clock-gating and then make a decision on which register to clock-gate. At netlist level this is very detailed and time consuming. In combinational clock-gating approach, this decision is approximated by applying clock-gating on registers of size above particular bit-width (e.g. 4- or 8-bit banks). Apart for this approximation another important step is to find out the common enables for different registers. This helps in putting a clock-gating cell near to the root of clock tree. In an HLS flow, netlist traversal may not always find out the common enable. For example, the same conditional assignment may exist in two different processes at HLS, which at RTL may represent two different state machine. If this common case can be extracted during HLS then even better clock-gated RTL can be generated. The algorithm presented in Chap. 10 takes care of such considerations.

The methodology presented in this chapter eliminates iterations requiring measurements of power consumption of each register before and after clock-gating is applied. It also shows, how the selection of appropriate register(s) would be performed *automatically* using simulation performed at transaction level. The extraction of enable provides two distinct advantages with respect to existing methodologies: (1) un-timed simulation can help in guiding the reduction results and (2) the reduction model specifically targeted for clock-gating gives an idea to the designer to find the possibilities of gating the logic. The proposed power model with technology specific parameters helps in finding this information. Another important point to note is that we can easily measure the impact of clock-gating, which helps us in getting the relative, yet accurate prediction on the savings opportunities.

Setup for our methodology includes C2R HLS tool [5], which takes ANSI C specification as input along with some constraints to generate RTL. These constraints help in finding an appropriate architecture for optimal timing and area. As illustrated in the Fig. 12.2, our methodology utilizes the HLS tool in two passes. In the first pass, tool generates cycle accurate transaction level (CATL) model in SystemC. In the generated SystemC simulation model, we embed the "relative" power model as discussed in the Sect. 12.2. This SystemC CATL model is simulated and attached to the untimed transaction level test vectors using generic payload models and a report is generated. For every register as discussed in the Sect. 12.2, Equations (12.4) and (12.9) are evaluated. If the inequality shown in these equations is not satisfied then a register is not a candidate for clock-gating. After this step, all the registers requiring the same enable signal are sorted.

Once the selection of registers is done, the HLS tool is utilized in the second pass to generate the RTL. For this purpose, we insert the clock-gating directives corresponding to the selected registers in the source code. These directives are macros

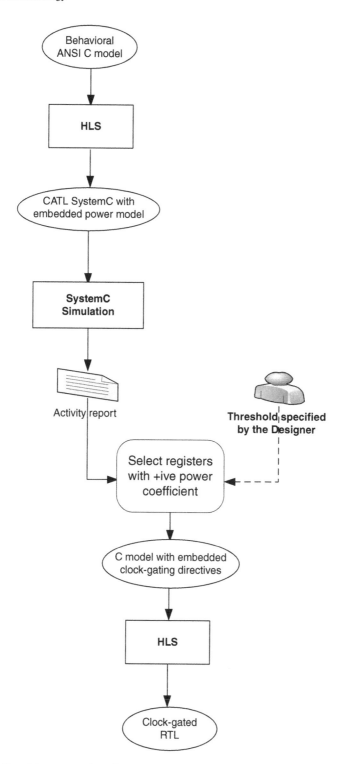

Fig. 12.2 Pictorial representation of our methodology

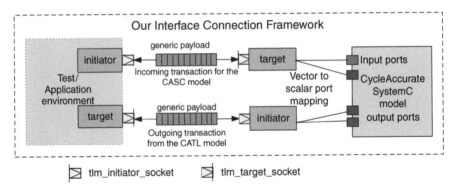

Fig. 12.3 Detailed view of system-level interconnect for auto-generated CATL models with transaction level testbench

such as clock_gating(ON) and not visible during the simulation performed in native C environment. Methodology described in the Chap. 10, shows how directives can be included in the C source code to handle various granularities of clock-gating. This auto-generated RTL contains clock-gating for the selected registers and will provide dynamic power savings as predicted. We have experimentally validated the accuracy of these predictions as reported later in this chapter. In our methodology, there is an optional input from the designer. Designer can increase the threshold if he/she suspects that a typical application scenario would exhibit lesser "savings duty cycle." Registers showing larger "savings duty cycle" than the threshold value are clock-gated. By default, this threshold is set to zero.

One such scenario where designer may want to put a high threshold value is the presence of a lot of local enables in a block with very high duty cycle. It suggests that locally there might not be very high savings existing in the design by applying clock-gating. Alternatively, it gives a handle to the designer to perform a what-if analysis on power savings to choose a coarse or fine grain clock-gating. Generally in block level clock gating the duty cycle of clock is reduced, now further inserting an icgc in clock tree path might not be desirable unless the local gating provides sufficient savings.

12.4.1 TLM Interface for the Generated CATL SystemC Model

Figure 12.3 shows how the CATL is attached to other TLM using simple sockets and generic payload. Generic payload is a data format defined in TLM2.0 standard for transport objects [117]. It contains data, address information, and typical attributes for memory, etc. It also has a response status field, so that recipient and caller can communicate status information. Our environment provides a communication medium between two parts, first is the test/application development platform where different application or test generators are running. Second is the CATL SystemC model of the design (DUT). Both test and design side works as recipient and caller while sending input and output for the design. For sockets, they are termed as initiator and target. Initiator initiates the transaction. We can see both the test and design having initiator and target sockets. Initiator from the test is connected to the target for the design (generated from the HLS tool). The ports on the cycle accurate model are scalar while the communication through the generic payload is using vectored data.

To ease the connection between the generic payload and the ports of CATL, we remodel the ports of the CATL as a part of a vectored representation of scalar ports. In generic payload models, the data is in the form of array of bytes, hence we used the port connections with a byte view of system-level interconnect for auto-generated CATL models. On the application/testbench side we initiate a transaction using generic payload data type. This interface helps in running the CATL model in the test environment to obtain the simulation statistics.

Snippet 12 tlm_data shows one such data structure. We provide a header file that can be used with any design. The purpose of this header is to connect the input and output transactions to the generated CATL model of the design. In this header, we first create a struct data type for both input and output transactions (one can use array as well). In this struct, we have the information for the input and output transactions. In generic payload, the data is in the form of array of bytes; hence, we used the port connections with a byte granularity.

Here in this transaction the data pointer is set to the data which is used for transaction, length and streaming width of the transaction is also defined. With this information all the port connections can be automatically performed.

Snippets 13 and 14 shows the view from the user side. Some trivial details on the interface functions for input_data and transaction are not explained in detail. Interested readers may refer to SystemC TLM distribution.

12.5 Results

In our experiments, we have used production quality industry strength benchmarks. Our benchmarks include various application domain such as compression, decompression, security, DSP related design blocks. RTL/CATL description of the designs are auto-generated and there is no manual modification involved in any part of the design flow. CATL is functionally equivalent to the RTL, cycle-by-cycle. Every part of the design flow is automated using makefile utility. In all experiments, RTL power estimation was performed using PowerTheater [129]. RTL simulation was performed using Modelsim [104], the high-level synthesis was performed using C2R HLS tool [41]. In our analysis, all the benchmarks were synthesized using standard cells hence register power is the biggest component. We did not replace memory with registers. In case memory is used, similar savings can be obtained for the total remaining registers in the design.

Snippet 12 Details of our interface and data structure to facilitate vector to scalar mapping on the design side

```
#define DATASIZE 8096
// DATASIZE is set for vector mapping of TLM and
// struct data type, it also represents the size of the
// transaction
typedef struct
{
    uint8_t data[DATASIZE];
}tlm_data;

static tlm_data local_in_data;
static uint16_t local_in_len;
static tlm_data local_out_data;
static uint16_t local_out_len;

// interface function for input output connectivity
// with the test environment
void interface_function (tlm_data in_data, uint16_t
in_len, tlm_data out_data, uint16_t out_len)
{
    local_in_data = in_data;
    local_in_len = in_len;
    out_data = local_out_data;
    out_len = local_out_len;
}
```

Snippet 13 Setting up the transaction from test environment side

```
tlm :: tlm_generic_payload * test_trans;
trans− > set_data_ptr(test_out_data);
trans− > set_data_length(n);
trans− > set_streaming_width(n);
```

Snippet 14 Sending the transaction from the test side

```
local_out_data.data[0] = 6;
local_out_len = 1;
```

In the Table 12.1, column 1 represents the design name, column 2 represents the dynamic register power measured at RTL without clock-gating, column 3 represents the power measured after clock-gating, column 4 represents the % saving shown at the RTL, column 5 shows the % savings predicted by our relative power model without any technology specific information and column 6 (Err1) shows the % error in estimation and column 7 shows the error in measurement using our power model with technology specific information. On these benchmarks, worst case error of our prediction is 7%, but in most cases we are very close. Even without the information on standard cells the accuracy is off by just 11% in the worst case. This presents a case that such a power reduction model can be used to find the effectiveness

Table 12.1 Percentage power reduction summary on the tested designs

Design	P_{nocg} (uW)	P_{cg} (uW)	R_{save}%	P_{pm}%	Err1.%	Err2.%
Fibo.	785	467	46.75	47	0.25	5.03
Caesar	1,630	191	88.3	87.71	0.6	5.85
FIR	2,090	768	63.3	72.67	9.37	5.07
DES	36,500	2,350	93.4	96.5	3.1	0.1
Volt.	3,240	391	98.8	97.06	1.74	5
Gzip	1,74,000	30,100	82.7	94	11.3	7.2
Gunzip	85,200	18,000	78.9	85.85	5.95	0.61
Vit.	48,200	20,800	56.84	50.66	6.2	7.01

Table 12.2 Speedup with our approach

Design	Our Ana.(s)	RTL (s)	Spd.up	Area sav. %
Fibo.	TS	TS	TS	24
Caesar	TS	TS	TS	9.3
FIR	TS	TS	TS	8.3
DES	TS	TS	TS	.7
Volt.	TS	TS	TS	0.01
GZIP	17.27	5,153.37	298	0
Gunzip	25.4	77,922	3,067	5.1
Viterbi	2.3	1,886	820	0.01

of power reduction transformations applied at RTL or higher-level of abstractions. In this chapter, we show the effectiveness on only one type of power reduction i.e. clock-gating, but this framework can easily be extended for other power reduction techniques.

Table 12.2 shows the speed up of the register selection process using our method over methods involving RTL power estimation tools. Here column 2 shows the CPU time spent in our analysis, column 3 shows the CPU time spent in RTL analysis, column 4 shows the speedup obtained and column 5 represent the area savings obtained after applying the clock-gating as compared to non clock-gated designs. Note that TS represents cases where the time taken (too small to measure) were in the range of a few seconds and hence not significant. We only show the cases where the RTL estimation took at least 1,000 s and then compared against our analysis timings. Note that in the case of the Gunzip benchmark, speedup reached around 3,000 times. This can be attributed to two factors. First, the system level analysis is much faster than RTL power analysis (expected in the range of 100–1,000 times). Second, the size of the vcd was quite large and it was provided as zipped vcd. It required additional processing to unzip the vcd.

One may argue that a relative power model can be created at the RTL as well, but the majority of power estimation time was spent at RTL simulation and simulation dump analysis. This step can not be avoided for RTL activity/power estimation and hence a power model at the RT level will never be as fast as one at the transaction level.

Table 12.3 Number of registers not to clock gate under various thresholds

Design	Auto	15%	25%	Total regs
Fibo.	25	59	61	91
Caesar	8	10	12	194
FIR	0	88	90	243
DES	32	51	125	4,326
Volt.	32	32	34	2,048
GZIP	208	382	459	17,909
Gunzip	280	745	1,077	2,266

As mentioned earlier, our default threshold is 0%, which means, that any register that is not 100% active will be considered a clock gating candidate. However, clock gating all or close to 100% registers may incur area overhead and timing closure issues. Also an icgc will be fed with clock port with 100% activity. Only the generated clock in the clock-tree would be having a lesser activity but the cost of including icgc may overcome the savings. We experimentally verified this claim. We ran some experiment on impacts of the choice of the threshold value such as 0%, 15%, and 25%. In the following, we present results for distinct choices for thresholds at 0%, 15%, and 25%. Then we calculated how much power can really be saved in each scenario.

Table 12.3 shows how many registers in the each design show negative values for (12.4) discussed in the Sect. 5.1, with each of the threshold settings. First, we saw that algorithm by default tries to find out quite a few registers that should not be clock-gated. Once, we increase the threshold, number of registers needs to be clock-gated decreases. Earlier, from the Table 12.1, we can see that on an average every case design is saving power but finding out the registers not saving power is extremely difficult. Our methodology very efficiently solves this problem.

Our experiments show that additional 2–10% of total register power could have been saved by applying clock-gating selectively. The fourth column of the Table 12.3 shows total number of registers in the designs. The table shows that up to 45% of the registers would not need clock-gating to save the power for Gzip design. Also for the same designs the number of icgc's that can be saved to insert in the clock-tree ranges upto 260 cells. This is a huge saving in terms of optimizing for the clock tree. This saving comes from the fact that our methodology can effectively find out the places to insert clock-gates only when it is needed.

Interestingly, we did not find much area overhead due to clock gating of selective registers. This can be attributed to two reasons: (a) we do a common enable analysis (by a traversal through CDFG of the design) to make sure that clock-gating logic is introduced for more than one register, which helps in reducing more multiplexers and (b) including a threshold helps in selectively introducing clock-gating. We found that a lot of single bit registers are not a good candidates for clock-gating; some times even 8-bit registers might not require clock-gating. However, we found that register banks of size 16 bits or above show a lot of power savings using clock-gating without exceptions.

Chapter 13
Conclusions

Power estimation and reduction have become primary concerns for hardware design. Increasing adoption of HLS tools in design flow is necessitating the proposition of power estimation and reduction methodologies at higher levels of abstraction. This thesis developed various techniques in these two areas. We observed that for power estimation accuracy and efficiency are the two areas of concern. In most of our approaches, we put an effort in providing a rationale behind the idea and then backed it with experimental results. We also proposed power reduction at high abstraction levels using various granularities of control for clock-gating based power reduction. Finally, we show how to utilize power estimation to guide power reduction. On the power estimation side, we present three different approaches (a) reusing the RTL power estimation frameworks at higher level, (b) providing characterization based power models to facilitate the power estimation at high-level and (c) utilize various verification collaterals such as assertions and properties to speedup power estimation and thereby the design time.

Chapter 5 presented characterization based power estimation methodology, utilizing GEZEL co-simulation environment. In this environment, hardware modeled as FSMD in GEZEL language can be co-simulated with existing processor models such as ARM. The power model proposed in this case study is a linear regression model, which is trained through various statistical test data. This regression based power model utilizes various states of FSMD as regression variables compared to input and output based characterization used in previous approaches. It gives a better visibility of power consumption because this approach helps in relating activity of the states of FSMD to power consumption. This power model is verified on a variety of benchmarks modeled in GEZEL and co-simulated with arm processor model. Further, the work is aided by working on different technology nodes (90 nm and 130 nm) and an elaborate discussion is provided on how to improve the estimation process when lower level power reduction techniques such as clock-gating, power-gating etc. are applied to the design. These reductions are introduced after the RTL

S. Ahuja et al., *Low Power Design with High-Level Power Estimation and Power-Aware Synthesis*, DOI 10.1007/978-1-4614-0872-7_13, © Springer Science+Business Media, LLC 2012

design stage, so it is important to know the impact of such techniques on power consumption at high-level. Error in the analysis is found to be less than 10% for the benchmarks used in our experiments.

Chapter 6 provides a proof-of-concept for high-level power estimation methodology for FPGA-based designs. The linear regression technique was adapted to develop generic power estimation model for multiple IPs. Both the I/O data patterns as well as the resource utilization of the designs were considered for the model by linearizing the power equation. It was first shown that, with a reasonably large data samples, one can get a good accuracy (less than 4% error, in out case), when there is only variation in data pattern. For the scenarios where the new IPs are introduced, one can either have a single regression model for all the IPs with relatively larger errors or can have multiple models for the IPs of different size and data pattern.

This work can be extended in multiple ways. Instead of considering XPower data for reference model, one could directly measure the board power consumption. This would be more accurate. Also, the static power component can be taken into account. Similar techniques can be used to measure the power consumption during downloading and configuring the bitstreams on the FPGAs. This would be important for the designs which make use of partially reconfigurable platforms.

Chapter 7 presented a high level synthesis based approach for solving performance constrained Hardware/software co-design problems. We identified key bottlenecks in the traditional approach and split the design space exploration loop into quick frequent and slow infrequent parts. The abstraction level of quick frequent loop was raised to behavioral level. The exploration at this level was done in C environment. We were able to perform coprocessor design space exploration at multiple levels – in terms of making coprocessor selection, making platform dependent trade offs for FSL, DMA, etc. considering Xilnx Spartan-3e as target and finally making coprocessor microarchitecture variation in the AES example. Considering the software-only version as the baseline for each of the IP, we provided relative speedups for all of them. The area of synthesized hardware was provided in terms of slice count. The power estimation was carried out for entire system to contrast the hardware and software versions. Results show that HLS based methods are suitable for design space exploration at many different levels.

Chapter 8 presented the idea of reusing RTL power estimation frameworks in HLS based design methodologies. This idea comes from the fact that in an HLS enabled design flow, RTL is tool generated and the biggest bottleneck in power estimation is the processing of simulation dump (vcd/fsdb). To reduce the time spent by the tool in performing power estimation, we showed how at higher levels speedup could be achieved by using system-level simulation and RT level design details. We experimentally show on a variety of benchmarks that the range of speedup in power estimation process might be as much as 15 times over RTL power estimation techniques. The error or loss of accuracy for the proposed methodology was less than 10% with respect to RTL. The benchmarks experimented for this methodology include a processor model (VeSPA), DSP algorithms such as (FFT, FIR) etc.

Power estimation methodologies/tools may be of no use if it is not aided with representative simulation scenarios. Chapter 9 presented a methodology to utilize verification collaterals to enhance the accuracy of power estimation at higher levels.

We used an Esterel based high level modeling framework to model control intensive designs such as a power state machine controller. We presented a methodology to verify various properties extracted from the specification using the formal verification framework inbuilt into the Esterel Studio. We provide guidance on how to write negative properties to help in creating counter cases. This guidance helps us in providing cases for reachability of a state, transitions from a state and staying in particular state of the controller using invariants. These counter cases can be synthesized to test-vectors, and data vectors are used as random test vectors while control test cases are created by utilizing properties. These comprehensive sets of test vectors provide the behavior of the controller of the design for various different states of the design at higher level. The model used in our study has four different states of a processor i.e. active, idle, sleep and deep sleep. We measure the power for each state and valid transitions and finally utilize these numbers during the simulation of the model.

On power reduction side, we proposed various flavors of clock-gating based power reduction. Such a reduction techniques is extensively used during logic synthesis. Overarching idea for our approach was to facilitate these reductions at higher-level and check its efficacy at that level itself. In that direction our proposal contains the following: (a) enabling clock-gating from ANSI C description itself, (b) enabling sequential clock-gating and finding a relationship of various registers at high-level and (c) enabling selective clock-gating, where guidance is provided by high-level power reduction model.

Chapter 10 presented a power reduction technique based on clock-gating from high level model description. We use a C2R HLS tool for our cases studies and show how from ANSI C description a clock-gated RTL can be auto-generated. We show, how various granularities of clock-gating can be affected from the high level description itself. This helps in finding out the opportunities, which lower level tools cannot achieve such as block-level clock-gating. In an HLS context, this is important because a lot of the block-level clock-gating decisions are done at the system-level. However, the design complexity is increasing and a lot of designers have started using the HLS to design the hardware blocks. We showed how various pragmas related to clock-gating can be introduced for an HLS tool to recognize them. We also show how to correctly capture the designer's intent on power management using clock-gating with these pragmas. Our studies also show the utilization of Integrated Clock-Gating Cell (ICGC) in generated RTL. ICGC is generally provided by the library vendors. Our proposed algorithm considers finding out the common enables for the registers so that across the blocks much lesser number of ICGCs are inserted on the clock-path. The other indirect benefit of such an approach is the introduction of ICGC closer to the root of the clock, otherwise there would be lot more ICGC's towards the leaf of the clock-tree, causing more power wastage in the clock tree.

To extend the clock-gating based power reduction, Chap. 11 showed how sequential clock-gating opportunities can be identified at higher abstraction levels. Generally the difficulty to implement sequential clock-gating comes from two facts (1) designers may inadvertently change the intended behavior of a design while trying out such aggressive reductions and (2) if tools are used to facilitate such an optimization, they generally work at netlist level and try to find out the observability don't care or stability conditions. These conditions suggest that a change in the value of register may not show impact on other register's output. Such an exploration at high-level improves the design cycle greatly. In this regard, we proposed a model checking based technique, which utilizes temporal properties to facilitate such a power reduction. It is shown that dependence of registers with each other can be captured in a property and then supplied to a model checker. If the model-checker passes that property then those registers become the possible candidates for sequential clock-gating. These conditions are put as "always true" properties, which is a strong condition but this helps in reducing the chances of facing bugs later in the design stage. This approach addresses the two problems mentioned earlier: (1) enabling ease of verification and (2) since the relationship between registers is obtained for a behavioral model, there is no need to go to the netlist level. This work is further extended on an HLS tool. Our experimental results show more power savings than conventional clock-gating based power reduction techniques.

Chapter 12 presented a methodology to automatically find the clock-gating opportunities by performing a system-level simulation of a design. To find such an opportunity, a system level power estimation model is used. It utilizes certain properties of clock-gating logic. It shows how technology dependent and independent models help in measuring the suitability of clock-gating applied at higher-level. These models help in applying the clock-gating at appropriate places in the design. The power model is integrated with untimed transaction level test/application environment. Our experimental results show 2–3 orders of magnitude faster analysis can be performed as compared to the state of the art RTL power estimation techniques. This facilitates analysis of more power reduction opportunities and helps in putting the appropriate granularity of clock-gating. This can also serve as a basis for applying estimation guided reduction for other optimizations such as sequential clock-gating, operand isolation, memory gating, etc. Currently, most of the approaches rely on putting the reduction algorithm based on architecture or designer knowledge, while this approach distinguishes itself by finding possible power savings by utilizing fast and accurate power reduction models.

This framework could be extended for a coprocessor design where a multi-criteria objective such as power, area and speed of co-processor can be taken care of, at the behavioral description itself. This will further reduce the need to traverse the lengthy outer design loop explained in the chapter.

13.1 Future Work

This thesis presented power estimation and reduction techniques for high level synthesis based design frameworks. There are many possible future directions, where the work presented here will be useful:

1. Power estimation framework presented in Chap. 5 can be extended in many ways. One of the possible extensions is to provide model for leakage power. Regression

model presented in this thesis considers constant leakage power. However, for lower technology nodes this relationship is not constant, thus an effort is required to improve the developed model. Second possibility is to extend the regression model for multiple IPs, that is one model to measure power consumption of multiple IPs. Such a model will make the estimation task very easy and requires minimal changes for new architectures and technology nodes.

2. Estimation framework utilizing verification collaterals presented in Chap. 9 can be improved. In our framework, we used random vectors to capture various simulation scenarios in a particular state of the design. This method can be augmented with test compaction techniques. Currently, exhaustive set of vectors are required to capture possible power consumption scenarios. A good compaction mechanism can help in capturing various power consumption patterns by using lesser simulation vectors.

3. We presented dynamic power reduction techniques at high-level (mainly clock-gating and its extensions). One of our approaches has shown enabling clock-gating from ANSI C description. This approach can be extended for some of the dynamic and static power reduction techniques such as operand-isolation, sequential clock-gating, power-gating. This will help in reducing the power consumption of design using multiple techniques while remaining at high-level.

4. In this book, we presented how a power reduction model guides HLS to reduce power consumption of a design. In the recent past, some trends suggest power wastage caused by overdoing the optimization. Our experimental results also confirm such trends. There are several possible extensions that can be applied to this power model, such as trade-offs for using operand-isolation, sequential clock-gating, memory-gating, power-gating, etc.

5. Formalization of the high-level synthesis framework helps us in controlling and getting better view of how design will look like. Existing frameworks such as C2R, used in this thesis, require manual refinements on the functional C model of the target design. These manual refinements are applied to convert the pure sequential code into a structural C code, which contains the information of the macro-architecture of the design. In the macro-architecture, generally designers aim to include necessary parallelism using threads or fork to explicitly bring out the concurrency opportunities in the sequential C-code. During the conversion process various minute details are inserted in the design description, which includes pipelining stages to achieve the desired throughput or necessary parallelism needed to meet the performance requirement, etc. A study on formalizing various intermediate stages to generate the control-data-flow graph (CDFG) of the input macro-architecture will be very useful. It will help in checking if the refinements are preserving essential properties. Also the formalism of the refinements may help in finding more power saving opportunities during HLS.

6. Finally, our approaches were not extensively targeted for hardware software co-design purposes. It will be interesting to see if these power reduction approaches can be extended in performing hardware software partitioning to avail power aware designing. A robust power model may help in finding out opportunities to save power for FPGA based SoC designs requiring run-time or partial reconfiguration.

References

1. A Practical Guide to Low Power Design. http://www.powerforward.org/lp_guide/. Accessed 2008
2. A System for Sequential Synthesis and Verification. http://www.eecs.berkeley.edu/~alanmi/abc/. Accessed 2008
3. Agarwal N, Dimopoulos N (2008) High-level fsmd design and automated clock gating with codel. Can J Electr Comput Eng 33(1):31–38
4. Ahuja S, Gurumani ST, Spackman C, Shukla SK (2009) Hardware Coprocessor Synthesis from an Ansi c Specification, vol 26. IEEE Computer Society, Los Alamitos, CA, USA, pp 58–67
5. Ahuja S, Gurumani ST, Spackman C, Shukla SK (2009) Hardware coprocessor synthesis from an ansi c specification. IEEE Design Test Comput 26:58–67
6. Ahuja S, Mathaikutty DA, Lakshminarayana A, Shukla S (2009) Accurate power estimation of hardware co-processors using system level simulation. 22nd IEEE International SOC Conference, 9–11 Sept 2009, pp 399–402
7. Ahuja S, Mathaikutty DA, Lakshminarayana A, Shukla S (2009) Statistical regression based power models for co-processors for faster and accurate power estimation. 22nd IEEE International SOC Conference, 9–11 Sept 2009, pp 399–402
8. Ahuja S, Mathaikutty DA, Lakshminarayana A, Shukla SK (2009) SCoPE: statistical regression based power models for co-processors power estimation. J Low Power Electron 5(4):407–415
9. Ahuja S, Mathaikutty DA, Shukla S (2008) Applying verification collaterals for accurate power estimation. In: 9th International Workshop on Microprocessor Test and Verification (MTV), pp 61–66
10. Ahuja S, Mathaikutty DA, Shukla S, Dingankar A (2007) Assertion-based modal power estimation. In: 8th International Workshop on Microprocessor Test and Verification (MTV), Austin, TX, 5–6 Dec 2007, pp 3–7
11. Ahuja S, Mathaikutty DA, Singh G, Stetzer J, Shukla S, Dingankar A (2009) Power estimation methodology for a high-level synthesis framework. In: 10th International Symposium on Quality Electronics Design (ISQED), pp 541–546
12. Ahuja S, Shukla SK (2009) MCBCG: model checking based sequential clock-gating. In: IEEE International Workshop on High Level Design Validation and Test, 4–6 November 2009, pp 20–25
13. Ahuja S, Zhang W, Lakshminarayana A, Shukla SK (2010) A methodology for power aware high-level synthesis of co-processors from software algorithms. In: Proceedings of International VLSI Design Conference, India, January 2010, pp 282–287

S. Ahuja et al., *Low Power Design with High-Level Power Estimation and Power-Aware Synthesis*, DOI 10.1007/978-1-4614-0872-7,
© Springer Science+Business Media, LLC 2012

14. Ahuja S, Zhang W, Shukla SK (2010) System level simulation guided approach to improve the efficacy of clock-gating. In: 15th IEEE International Workshop on High Level Design Validation and Test, June 2010, pp 9–16
15. Babighian P, Benini L, Macii E (2004) A scalable ODC-based algorithm for RTL insertion of gated clocks . Design Automation and Test in Europe, (DATE'04), Paris, France, February 2004, pp 720–721
16. Banga M, Chandrasekar M, Fang L, Hsiao MS (2008) Guided test generation for isolation and detection of embedded trojans in ics. pp 363–366
17. Banga M, Hsiao MS (2008) A region based approach for the identification of hardware trojans. In: HOST, pp 40–47
18. Banga M, Hsiao MS (2009) A novel sustained vector technique for the detection of hardware trojans. In: Proceedings of the International Conference on VLSI Design, pp 327–332
19. Banga M, Hsiao MS (2009) Vitamin: Voltage inversion technique to ascertain malicious insertions in ics. In: HOST, pp. 104–107
20. Bansal N, Lahiri K, Raghunathan A, Chakradhar ST (2005) Power monitors: a framework for system-level power estimation using heterogeneous power models. In: Proceedings of the 18th International Conference on VLSI Design (VLSID'05)
21. Benini L, Bruni D, Chinosi M, Silvano C, Zaccaria V, Zafalon R (2001) A power modeling and estimation framework for vliw-based embedded systems. In: Proceedings of International Workshop-Power and Timing Modeling, Optimization and Simulation, PATMOS'01
22. Bergamaschi RA, Shin Y, Dhanwada N, Bhattacharya S, Dougherty WE, Nair I, Darringer J, Paliwal S (2003) Seas: a system for early analysis of socs. In: CODES+ISSS '03: Proceedings of the 1st IEEE/ACM/IFIP International Conference on Hardware/Software Codesign and System Synthesis, ACM, New York, NY, USA, pp 150–155
23. Berry G (2000) The foundations of ESTEREL. In: Proof, Language and Interaction: Essays in Honour of Robin Milner, pp 425–454
24. Berry G (1992) Esterel on hardware. In: Philosophical Transactions of the Royal Society of London (Series A, 339), April 1992, pp 87–104
25. Berry G, Gonthier G (1992) The esterel synchronous programming language: design, semantics, implementation. Sci Comput Program 19(2):87–152
26. Bingham J, Erickson J, Singh G, Andersen F (2009) Industrial strength refinement checking. In: Formal Methods in Computer Aided Design (FMCAD), Austin, Texas, USA, November 2009
27. Bluespec inc. http://www.bluespec.com/. Bluespec Compiler. Accessed 2009
28. Bluespec Inc. Bluespec. http://bluespec.com/. Accessed 2009
29. Bluespec Inc. Bluespec. http://bluespec.com/. Accessed 2009
30. Bogliolo A, Benini L, De Micheli G (1997) Adaptive least mean square behavioral power modeling. Conference on Design Automation and Test in Europe
31. Bogliolo A, Benini L, De Micheli G (1998) Characterization-free behavioral power modeling. Conference on Design Automation and Test in Europe
32. Bogliolo A, Colonescu I, Macii E, Poncino M (2001) An rtl power estimation tool with on-line model building capabilities. In: International Workshop on Power Timing Modeling, Optimization Simulation Yverdon-les-Bains, Switzerland
33. Bogliolo A, Benini L, De Micheli G (2000a) Regression-Based rtl Power Modeling, vol 5. ACM, New York, NY, USA, pp 337–372
34. Bogliolo A, Benini L, De Micheli G (2000b) Regression-based rtl power modeling. ACM Trans Des Autom Electron Syst 5(3):337–372
35. Bona A, Sami M, Sciuto D, Silvano C, Zaccaria V, Zafalon R (2005) Reducing the complexity of instruction-level power models for vliw processors. In: Design Automation for Embedded Systems
36. Brandt J, Schneider K, Ahuja S, Shukla SK (2010) The model checking view to clock gating and operand isolation. In: Proceedings of ACSD, 21–25 June 2010, pp 181–190
37. Cadence Design Systems. C-to-silcon Compiler. http://www.cadence.com/products/sd/silicon_compiler/pages/default.aspx. Accessed 2009

38. Cadence Inc. The Cadence SMV model checker. http://www.kenmcmil.com/smv.html. Accessed 2009
39. Cai L, Gajski D (2003) Transaction level modeling: an overview. IEEE/ACM/IFIP International Conference on Hardware/Software Codesign and System Synthesis
40. Caldari M, Conti M, Coppola M, Crippa P, Orcioni S, Pieralisi L, Turchetti C (2003) System-level power analysis methodology applied to the amba ahb bus. In: Proceedings of the conference on Design, Automation and Test in Europe (DATE)
41. Cebatech Inc. C2R Compiler. http://www.cebatech.com. Accessed 2008
42. Celoxica Limited – Agility Compiler – Advanced Synthesis Technology For SystemC. http://www.celoxica.com. Accessed 2007
43. Celoxica. Handel-C Language Reference Manual RM-1003-4.0, 2003. http://www.celoxica.com. Accessed 2003
44. Celoxica. Handel-C Language Reference Manual RM-1003-4.0, 2003. http://www.celoxica.com. Accessed 2003
45. Chandrakasan AP, Potkonjak M, Mehra R, Rabaey J, Brodersen RW (1995) Optimizing Power Using Transformations. IEEE Trans Comput Aided Design Integr Circ Syst 14:12–31
46. Chang J-M, Pedram M (1999) Power Optimization and Synthesis at Behavioral and System Levels Using Formal Methods. Kluwer Academic Publishers, Dordecht
47. Chattopadhyay A, Geukes B, Kammler D, Witte EM, Schliebusch O, Ishebabi H, Leupers R, Ascheid G, Meyr H (2006) Automatic ADL-based operand isolation for embedded processors. In: Design Automation and Test in Europe (DATE'06)
48. Chen D, Cong J, Fan Y (2003) Low-power high-level synthesis for FPGA architectures. IEEE/ACM International Conference on Low Power Electronics and Design (ISLPED'03), August 2003, pp 134–139
49. Cong J, Liu B, Zhang Z (2009) Behavior-level observability don't-cares and application to low-power behavioral synthesis. In: ISLPED '09: Proceedings of the 14th ACM/IEEE International Symposium on Low Power Electronics and Design, ACM, New York, NY, USA, pp 139–144
50. Dal D, Kutagulla D, Nunez A, Mansouri N (2005) Power islands: a high-level synthesis technique for reducing spurious switching activity and leakage. In: 48th Midwest Symposium on Circuits and Systems, 2005, vol 2, August 2005, pp 1875–1879
51. Degalahal V, Tuan T (2005) Methodology for high level estimation of fpga power consumption. In: ASP-DAC '05: Proceedings of the 2005 Asia and South Pacific Design Automation Conference, ACM, New York, NY, USA, pp 657–660
52. Deng YZL, Sobti K, Chakrabarti C (2009) Accurate area, time and power models for fpga-based implementations. In: Journal of Signal Processing, 2009, pp 39–50
53. De Micheli G (1999) Hardware synthesis from C/C++ models. In: Design Automation and Test in Europe Conference and Exhibition 1999 (DATE'99), March 1999, pp 382–383
54. De Micheli G, Ku D, Mailhot F, Truong T (1990) Olympus Syn Syst 7:37–53
55. Domer R, Gerstlauer A, Gajski D (2001) SpecC Language Reference Manual, Version 2.0. March 2001
56. Edwards SA (2002) High-level synthesis from the synchronous language esterel. In: International Workshop of Logic and Synthesis, New Orleans, Louisiana (IWLS'02)
57. Edwards SA (2004) The challenges of hardware synthesis from C-like languages. In: International Workshop on Logic Synthesis, Temecula, California (IWLS'04), June 2004, pp 509–516
58. Emnett F, Biegel M (2000) Power reduction through RTL clock gating. In: Presented at the Synopsys Users Group (SNUG), San Jose, CA
59. ESPRESSO Project, IRISA. The Polychrony Toolset. http://www.irisa.fr/espresso/Polychrony. Accessed 2009
60. ESTEREL Technologies. Esterel Studio. http://www.esterel-technologies.com/products/esterel-studio/. Accessed 2007
61. Forte Design Systems. Cynthesizer. http://www.forteds.com. Accessed 2009

62. Forte Design Systems. Low Power Design flow. http://www.forteds.com/solutions/lowpower. asp. Accessed 2009

63. Fraer R, Kamhi G, Mhameed MK (2008) A new paradigm for synthesis and propagation of clock gating conditions. In: DAC '08: Proceedings of the 45th Annual Design Automation Conference, ACM, New York, NY, USA, pp 658–663

64. Frenkil J. Too Much of a Good Thing? http://chipdesignmag.com/display.php?articleId=3316. Accessed 2009

65. Gajski DD, Zhu J, Domer R, Gerstlauer A, Zhao S (2000) SpecC: Specification Language and Methodology. Kluwer Publications, Dordecht

66. GAUT. High-Level Synthesis Tool from C to RTL. http://wwwlabsticc.univ-ubs.fr/www-gaut/. Accessed 2010

67. Geilen M, Basten T, Theelen B, Otten R (2007) An algebra of pareto points. Fundam Inf 78(1):35–74

68. GEZEL reference manual. GEZEL Language Information. http://rijndael.ece.vt.edu/gezel2/index.php/Main_Page. Accessed 2010

69. GEZEL reference manual. GEZEL Language Information. http://rijndael.ece.vt.edu/gezel2/index.php/Main_Page. Accessed 2009

70. Ghenassia F (2006) Transaction-Level Modeling with Systemc: Tlm Concepts and Applications for Embedded Systems. Springer-Verlag New York, Inc., Secaucus, NJ, USA

71. Guernic PL, Borgue M, Gautier T, Marie C (1991) Programming real time applications with SIGNAL. Proc IEEE 79(9):1321–1335

72. Guo X, Chen Z, Schaumont P (2008) Energy and performance evaluation of an fpga-based soc platform with aes and present coprocessors. In: 8th international Workshop on Embedded Computer Systems: Architectures, Modeling, and Simulation, July 2008

73. Gupta S, Gupta RK, Dutt ND, Nicolau A (2004) SPARK: A Parallelizing Approach to the High-Level Synthesis of Digital Circuits. Kluwer Academic Publisher, Dordecht

74. Gupta S, Najm FN (1997) Power macromodeling for high level power estimation. In: Design Automation Conference

75. Gupta S, Najm FN (2003) Energy and peak-current per-cycle estimation at rtl. In: IEEE Transactions on VLSI Systems

76. Hoare CAR (1978) Communicating sequential processes. Commun ACM 21(8):666–677

77. Hsieh C-T, Wu Q, Ding C-S, Pedram M (1996) Statistical sampling and regression analysis for rt-level power evaluation. In: *ICCAD '96: Proceedings of the 1996 IEEE/ACM International Conference on Computer-Aided Design*, IEEE Computer Society, Washington, DC, USA, pp 583–588

78. Irani S, Singh G, Shukla SK, Gupta RK (2005) An overview of the competitive and adversarial approaches to designing dynamic power management strategies. J IEEE Trans VLSI Syst (TVLSI) 13(12):1349–1361

79. Jha NK, Shang L (2001) High-level power modeling of cplds and fpgas. In: ICCD '01: Proceedings of the International Conference on Computer Design: VLSI in Computers and Processors, IEEE Computer Society, Washington, DC, USA, p 46

80. Jiang T, Tang X, Banerjee P (2004) High level area, delay and power estimation for fpgas. In: FPGA '04: Proceedings of the 2004 ACM/SIGDA 12th International Symposium on Field Programmable Gate Arrays, ACM, New York, NY, USA, pp 249–249

81. JMP Description. JMP Data Analysis. http://www.jmp.com/software/jmp7/. Accessed 2009

82. Jose BA, Patel HD, Shukla SK, Talpin J-P (2009) Generating multi-threaded code from polychronous specifications. Electron Notes Theor Comput Sci 238(1):57–69

83. Jose BA, Pribble J, Shukla SK (2010) Faster embedded software synthesis using actor elimination techniques for polychronous formalism. In: 10th International Conference on Application of Concurrency to System Design (ACSD2010)

84. Jose BA, Pribble J, Stewart L, Shukla SK (2009) EmCodeSyn: a visual framework for multi-rate data flow specifications and code synthesis for embedded application. In: 12th IEEE Forum on Specification and Design Languages (FDL'09), September 2009, pp 1–6

85. Jose BA, Shukla SK (2010) An alternative polychronous model and synthesis methodology for model-driven embedded software. In: Proceedings of IEEE Asia and South Pacific Design Automation Conference (ASP-DAC 2010), January 2010, pp 13–18

86. Jose BA, Shukla SK (2010) MRICDF: a polychronous model for embedded software synthesis. In: Synthesis of Embedded Software – Frameworks and Methodologies for Correctness by Construction Software Design

87. Jose BA, Shukla SK, Patel HD, Talpin J-P (2008) On the deterministic multi-threaded software synthesis from polychronous specifications. In: 6th ACM/IEEE International Conference on Formal Methods and Models for Co-Design, 2008. MEMOCODE 2008, June 2008, pp 129–138

88. Jose BA, Xue B, Shukla SK (2009) An analysis of the composition of synchronous systems. Electron Notes Theor Comput Sci 245:69–84

89. Jose BA, Xue B, Shukla SK, Talpin J-P (2010) Programming models for multi-core embedded software. Multi-Core Embedded Systems

90. Khouri KS, Lakshminarayana G, Jha NK (1999) High-Level synthesis of low-power control-flow intensive circuits. In: IEEE Transactions on Computer-Aided Design (TCAD'99), 18 Dec 1999

91. Ku D, De Micheli G (1990) HardwareC – A Language for Hardware Design (Version 2.0). *Technical Report: CSL-TR-90-419, Stanford University.*

92. Kumar A, Bayoumi M (1999) Multiple voltage-based scheduling methodology for low-power in the high-level synthesis. IEEE Int Symp Circuits Syst 1:371–374

93. Lajolo M, Raghunathan A, Dey S, Lavagno L (2000) Efficient power co-estimation techniques for system-on-chip design. In: Design Automation and Test in Europe (DATE)

94. Lakshminarayana A, Ahuja S, Shukla S (2010) Coprocessor design space exploration using high level synthesis. In: 10th International Symposium on Quality Electronics Design (ISQED), March 2010, pp 879–884

95. Lakshminarayana G, Raghunathan A, Jha NK, Dey S (1998) A power management methodology for high-level synthesis. International Conference on VLSI Design, January 1998, pp 24–29

96. Lee EA, Messerschmitt DG (1989) Static scheduling of synchronous data flow programs for digital signal processing. IEEE transactions on Computers, 1987, pp 237–248

97. Li H, Bhunia S, Chen Y, Vijaykumar TN, Roy K (2003) Deterministic clock gating for microprocessor power reduction. In: HPCA '03: Proceedings of the 9th International Symposium on High-Performance Computer Architecture, IEEE Computer Society, Washington, DC, USA, p 113

98. Lilja DJ, Saptnekar, SS Designing Digital Computer Systems with Verilog. http://www.arctic.umn.edu/vespa/. Accessed 2008

99. Liu X, Papaefthymiou MC (2002) A markov chain sequence generator for power macro-modeling. In: ICCAD '02: Proceedings of the 2002 IEEE/ACM International Conference on Computer-Aided Design, ACM, New York, NY, USA, pp 404–411

100. Mehta H, Owens RM, Irwin MJ (1996) Energy characterization based on clustering. In: Design Automation Conference

101. Mentor Graphics Corp. Catapult Synthesis. www.mentor.com/products/esl/high_level_synthesis/catapult_synthesis. Accessed 2009

102. Mentor Graphics Inc. Advanced Clock Gating Techniques in Catapult C Synthesis. http://www.techonline.com/learning/techpaper/218400046. Accessed 2009

103. Mentor Graphics Inc. RTL Simulation. http://www.model.com/. Accessed 2009

104. Mentor Graphics Inc. RTL Simulation. http://www.model.com/. Accessed 2009

105. Mitch Dale. The Power of RTL Clock-gating. http://www.chipdesignmag.com/display.php?articleId=915. Accessed 2009

106. Mohanty SP, Ranganathan N (2003) A framework for energy and transient power reduction during behavioral synthesis. In: Proceedings of the International Conference on VLSI Design, pp 539–545

107. Mohanty SP, Ranganathan N, Chappidi SK (2004) ILP models for energy and transient power minimization during behavioral synthesis. In: Proceedings of the International Conference on VLSI Design, p 745
108. Monteiro J, Devadas S, Ashar P, Mauskar A (1996) Scheduling techniques to enable power management. In: Proceedings of Design Automatin Conference, June 1996, pp 349–352
109. Moore GE (1965) Cramming more components onto integrated circuits. Electronics 38(8)
110. Munch M, Wurth B, Mehra R, Sproch J, Wehn N (2000) Automating RT-level operand isolation to minimize power consumption in datapaths. In: Design Automation and Test in Europe (DATE'00)
111. Murugavel AK, Ranganathan N (2003) A game-theoretic approach for binding in behavioral synthesis. In: 16th International Conference on VLSI Design, 2003, January 2003, pp 452–458
112. Namballa R, Ranganathan N, Ejnioui A (2004) Control and data flow graph extraction for high-level synthesis. IEEE Computer Society Annual Symposium on VLSI, 0:187
113. Negri L, Chiarini A (2006) Power simulation of communication protocols with statec. Applications of Specification and Design Languages for SoCs. Springer, Netherlands, pp 277–294
114. Nemani M, Najm FN (1996) Towards a high-level power estimation capability. IEEE Trans Comput Aided Des Integr Circuits Syst 15:588-598
115. Ohashi M et al. (2002) A 27mhz 11.1mw mpeg-4 video decoder lsi for mobile application. ISSCC Digest of Technical Papers, pp 366–367
116. OSCI Group. SystemC Website. http://www.systemc.org/. Accessed 2009
117. OSCI Group. TLM 2.0. http://www.systemc.org/downloads/standards/tlm20/. Accessed 2009
118. Patel H, Shukla S Power State Machine Implementation. http://fermat.ece.vt.edu/EWD/Power_State_Machine.htm. Accessed 2009
119. Potlapally NR, Raghunathan A, Lakshminarayana G, Hsiao M, Chakradhar ST (2001) Accurate power macro-modeling techniques for complex rtl components. In: Proceedings of International Conference on VLSI Design Bangalore, India, Jan 2001, p 235
120. Power Compiler, Synpsys Inc. http://www.synopsys.com. Accessed 2009
121. PowerPlay Early Power Estimator, Altera Inc. http://www.altera.com/literature/hb/qts/qts_qii53006.pdf. Accessed 2009
122. Raghunathan A, Jha NK (1994) Behavioral synthesis for low power. In: ICCAD'94, pp 318–322
123. Raghunathan A, Jha NK, Dey S (1998) High-Level Power Analysis And Optimization. Kluwer Academic Publishers, Dordecht
124. Raghunathan V, Ravi S, Raghunathan A, Lakshminarayana G (2001) Transient power management through high level synthesis. In: Proceedings of the ICCAD, pp 545–552
125. Ramesh J et al. Clock Gating for Power Optimization in ASIC. www.islped.org/X2008/Jairam.pdf
126. Russell J, Jacome M (1998) Software power estimation and optimization for high performance, 32-bit embedded processors. In: Proceedings of ICCD
127. Schneier B (1996) Applied Cryptography. Wiley, NY, USA
128. Sequence Design Inc. RTL Power Management. http://sequencedesign.com/solutions/powertheater.php. Accessed 2009
129. Sequence Design Inc. (2007) RTL Power Estimation. In: PowerTheater User guide, release 2007.2
130. Shang L, Kaviani AS, Bathala K (2002) Dynamic power consumption in virtex ii fpga family. In: FPGA '02: Proceedings of the 2002 ACM/SIGDA Tenth International Symposium on Field-Programmable Gate Arrays, ACM, New York, NY, USA, pp 157–164
131. Shin H, Lee C (2007) Operation mode based high-level switching activity analysis for power estimation of digital circuits. IEICE Trans Commun E90-B(7):1826–1834
132. Shiue WT (2000) High level synthesis for peak power minimization using ilp. In: Proceedings of the IEEE International Conference on ASSAP, pp 103–112

133. Shiue W-T, Chakrabarti C (2000) ILP-based scheme for low power scheduling and resource binding. In: IEEE International Symposium on Circuits and Systems (ISCAS'00), vol 3. Geneva, Switzerland, May 2000, pp 279–282

134. Shiue W-T, Chakrabarti C (2000) Low-power scheduling with resources operating at multiple voltages. IEEE Trans Circuits Syst - II: Analog Digital Signal Process 47(6):536–543

135. Singh G (2008) Optimization and Verification Techniques for Hardware Synthesis from Concurrent Action-Oriented Specifications. In: Phd dissertation Virginia Tech, 2008

136. Singh G, Gupta S, Shukla SK, Gupta R (2006) Chapter 11 of CRC Handbook on EDA for IC System Design, Verification and Testing, Chapter: Parallelizing High-Level Synthesis: A Code Transformational Approach to High-Level Synthesis. CRC, Taylor and Francis Group, Boca Raton

137. Singh G, Schwartz JB, Ahuja S, Shukla SK (2007) Techniques for power-aware hardware synthesis from concurrent action oriented specifications. J Low Power Electron (JOLPE) 3(2):156–166

138. Singh G, Schwartz JB, Shukla SK (2009) A formally verified peak-power reduction technique for hardware synthesis from concurrent action-oriented specifications. J Low Power Electron (JOLPE) 5(2):135–144

139. Singh G, Shukla SK (2006) Low-power hardware synthesis from TRS-based specifications. Fourth ACM and IEEE International Conference on Formal Methods and Models for Codesign (MEMOCODE'06), Napa Valley, CA, USA, July 2006, pp 49–58

140. Singh G, Shukla SK (2007a) Algorithms for low power hardware synthesis from CAOS – concurrent action oriented specifications. In: Special Issue of International Journal of Embedded Systems on Power/Energy/Thermal topics (IJES'07)

141. Singh G, Shukla SK (2007b) Model checking bluespec specified hardware designs. In: 8th International Workshop on Microprocessor Test and Verification (MTV), Austin, TX, USA, December 2007

142. Singh G, Shukla SK (2008) Verifying compiler based refinement of Bluespec specifications using the SPIN model checker. In: 15th International SPIN Workshop on Model Checking of Software (SPIN), Los Angeles, CA, USA, August 2008, pp 250–269

143. Singh G, Ravi SS, Ahuja S, Shukla SK (2007) Complexity of scheduling in synthesizing hardware from concurrent action oriented specifications. In: Power-aware Computing Systems, Dagstuhl Seminar Proceedings 07041, Dagstuhl, Germany

144. Stine JE, Grad J, Castellanos I, Blank J, Dave V, Prakash M, Iliev N, Jachimiec N (2005) A framework for high-level synthesis of system-on-chip designs. In: MSE '05: Proceedings of the 2005 IEEE International Conference on Microelectronic Systems Education (MSE'05), IEEE Computer Society, Washington, DC, USA, pp 67–68

145. Stroud CE, Munoz RR, Pierce DA (1988) Behavioral model synthesis with cones. IEEE Design Test Comput 5:22–30

146. Sullivan C, Wilson A, Chappell S (2004) Using c based logic synthesis to bridge the productivity gap. Asia and South Pacific Design Automation Conference, IEEE Press, New York, January 2004, pp 349–354

147. Synfora. PICO Express. http://www.synfora.com/. Accessed 2008

148. Synopsys Inc. Design for Low Power. http://www.synopsys.com/products/solutions/galaxy/power/power.html

149. Synopsys Inc. VCS Comprehensive RTL Verification Solution. http://www.synopsys.com/vcs/. Accessed 2009

150. Tiwari V, Malik S, Wolfe A, Lee M (1996) Instruction level power analysis and optimization of software. J VLSI Signal Process Syst 13:223–238

151. Tiwari V, Malik S, Wolfe A (1994) Power analysis of embedded software: a first step towards software power minimization. In: *ICCAD '94: Proceedings of the 1994 IEEE/ACM International Conference on Computer-aided Design*, IEEE Computer Society Press, Los Alamitos, CA, USA, pp 384–390

152. Uchida J, Togawa N, Yanagisawa M, Ohtsuki T (2004) A thread partitioning algorithm in low power high-level synthesis. Asia and South Pacific Design Automation Conference (ASP-DAC'04), January 2004, pp 74–79

153. Unified Modeling Language. http://www.uml.org/

154. Verification Interacting with Synthesis (VIS). http://vlsi.colorado.edu/~vis/. Accessed 2009

155. Vahid F (2006) Digital Design. Wiley, New York

156. Wakabayashi K (1999) C-based synthesis experiences with a behavior synthesizer, "Cyber". Design Automation and Test in Europe Conference and Exhibition 1999 (DATE'99), March 1999, pp 390–393

157. Weste NHE, Harris D, Banerjee A (2006) CMOS VLSI DESIGN: A Circuits and Systems Perspective. Pearson Education

158. Wolf W (2002) Household Hints for Embedded Systems Designers. IEEE Computer Society Press, Silver Spring, MD

159. Xilinx. Xilinx Power Estimator. http://www.xilinx.com/products/design_resources/power_central/. Accessed 2008

160. Xilinx. XPower Analyzer. http://www.xilinx.com/products/design_tools/logic_design/verification/xpower.htm. Accessed 2009

161. Xilinx Inc. Spartan-3E Starter Kit. http://www.xilinx.com/products/devkits/HW-SPAR3E-SK-US-G.htm. Accessed 2009

162. Xilnx Inc. MicroBlaze Softcore Processor Core. http://www.xilinx.com/tools/microblaze.htm

163. XPower Estimator (XPE), Xilinx Inc. http://www.xilinx.com/products/design_resources/power_central/. Accessed 2009

164. Zhong L, Ravi S, Raghunathan A, Jha NK (2004) Power estimation for cycle-accurate functional descriptions of hardware. International Conference on Computer Aided Design, 2004. ICCAD-2004